Business Writing For Everyday Use

SAMPLE
LETTERS
AND
MEMOS

FOR
BUILDERS, DEVELOPERS, AND REMODELERS

John A. Kilpatrick

HOME
BUILDER
PRESS

Home Builder Press®
National Association of Home Builders
15th and M Streets, NW
Washington, DC 20005

Sample Letters and Memos for Builders, Developers, and Remodelers: Business Writing for Everyday Use
ISBN 0-86718-374-8

© 1991 by Home Builder Press® of the National Association of Home Builders of the United States of America

Library of Congress Cataloging-in-Publication Data

Kilpatrick, John A., 1954 —
    Sample letters and memos for builders, developers, and remodelers : business writing for everyday use / John A. Kilpatrick.
        p.   cm.
    ISBN 0-86718-374-8
    1. Construction industry—Records and correspondence.   2. Commercial correspondence.   I. Title.
    HD9715.A2K543 1991
    651.7'5'02469—dc20   91-43255
                                                                                              CIP

For further information, please contact—

Home Builder Press®
National Association of Home Builders
15th and M Streets, NW
Washington, DC 20005
(800) 223-2665

12/91 HBP/United Book Press 2.5K

Contents

Chapter 7 Letters for Challenging and Special Situations 107

About the Author

John Aaron Kilpatrick holds an M.B.A. from the University of South Carolina (USC) and has served as controller of one of South Carolina's largest homebuilding and residential development firms. In that position, he was the first to computerize the business department of a large-scale homebuilding company in South Carolina. He is also president and broker-in-charge of Kilpatrick and Associates Realty, a development and consulting firm. He has taught finance and financial management at several colleges and universities. John Kilpatrick is currently a research associate on the staff of the Vice Provost for Research at USC, where he is also finishing his Ph.D. in finance.

Acknowledgments

Special thanks go to Bill Adams, Carol Smith, Bob Whitten, Mary DiCrescenzo, and Dawn Harris for their thoughtful review of the manuscript.

This book was produced under the general direction of Kent Colton, NAHB Executive Vice President, in association with NAHB staff members James E. Johnson, Jr., Staff Vice President, Operations and Information Services; Adrienne Ash, Assistant Staff Vice President, Publishing Services; Rosanne O'Connor, Director of Publications and Project Editor; Melissa Brown, Publications Editor; and David Rhodes, Art Director.

Disclaimer

The letters in this book are samples and are presented for illustrative purposes only. They do not and cannot apply to every situation. In some circumstances, a letter may give rise to legal obligations or ramifications and it may therefore be necessary to consult with your local attorney before signing such a letter. The National Association of Home Builders hereby disclaims any and all liability that may arise from the use of this book or its contents.

Chapter 1

Importance of Effective Business Writing

Why Put It in Writing?

Remember when you were a kid in the first or second grade and you played the whisper game? Ten or 20 kids sat in a circle and the teacher whispered a message into the first kid's ear. The message was always something like "Three cows named Marjorie, William, and Kevin are eating oats in the field next to the barn" or "Red-headed Vikings always sail west for the winter."

The first kid would whisper the message to the second kid who would whisper it to the third kid until the message made it back around to the teacher. By the end, "Three cows..." had become "Why do buzzards fly east to cross drawbridges?" or something like that. The point of the game was to show that verbal communication is often faulty so you should pass on messages carefully. That is also the underlying point of this book.

Besides presenting letters that you may adapt for your own use, this book emphasizes one clear and unmistakable point. People can easily garble, forget, or misunderstand verbal messages. As the saying goes, oral contracts are only as good as the paper they are written on.

As a builder you should put important things in writing for the following reasons:

- to avoid confusion
- to keep a permanent record of what you said
- to add emphasis to requests, orders, or policies
- to save time
- to project the professional image any business needs
- to ensure your financial success

Coordinating Oral and Written Communications

How many business people end the day with a pocket full of notes? These can range from a legal pad full of neatly organized decisions made during the day to scraps of paper in a pocket with semi-legible scribbles and scrawls. Some well-organized business people carry pocket tape recorders and dictate an endless stream of notes to themselves. At the other end of the spectrum, some people do not take any notes at all, hoping to rely on superhuman memories to carry them through.

Clearly, managing a business on a daily basis requires a great deal of on-the-spot decision making. Orders are issued, plans are changed, employees are assigned new duties, contracts are agreed upon, and a wide variety of promises are made and received concerning customers, vendors, and business associates. Few business people can take the time to put all of these decisions in writing.

However, you should put most business decisions in writing. If you have made an oral agreement, such as a special deal on a truckload of lumber or a bonus for a salesperson, then you need to follow this decision with a written letter or memo outlining the details of the agreement. That way, if any disagreements over the details arise, you can resolve them immediately.

As the first one to put an oral agreement in writing, you gain an added bonus. Oral agreements often have some points of disagreement, some points not covered, or some details to be worked out in the future. If you are the first to write the agreement down, you can set the tone and agenda for how the agreement will be put in place, and this can often work to your advantage.

How to Put It in Writing

To write effective business letters, provide just enough meat on the bones and leave it at that. You want your business letters to contain clear sentences of a reasonable length with little room for misunderstanding. However, you also want to avoid short, choppy phrasing that might sound cold and official. Business letters that work have a friendly flavor to them. Effective business writers will also favor ordinary words over technical words that might confuse the reader.

Whether you are writing a memo or letter, get to the point in the first paragraph and even the first sentence. The two most important sentences in any letter are the first and the last. The first sentence sets the agenda for the rest of the document, while the last sentence lets everyone know where you want to go from there.

The first sentence and paragraph of your letter also set the tone for the entire document. For example, "Joe's Building Company will order 10,000 board feet of 2x10s for the framing of the Norvell's new home" conveys a sense of action. It establishes a positive note for whatever follows. The sentence projects the vital image most businesses want.

Besides being forceful or weak, the opening sentence and paragraph of your letter can be either positive or negative. This aspect by itself conveys an important message. As an example, "I am pleased to announce that the sales for Joe's Builders grew by an impressive 5.5 percent last year, despite a recession in the housing market" leaves a positive impression about the company. Contrast that sentence with the following, more negative opening sentence: "I am sure you share my disappointment that sales for Joe's Builders grew by only 5.5 percent last year because of the recession in the housing market." Get the picture? Here are two sentences that say the same thing but set different tones for the letters that follow.

So, carefully word the opening sentence and paragraph of your letters to set the tone you wish to convey for the rest of the letter. In the same way, always end your letter on a positive, action-oriented note.

Often business letters try to persuade someone to do something. Many poor writers fall back on gimmicks or tricks to persuade others to act. Some of these gimmicks might include omitting ideas, trying to hide ideas under a blanket of pretty descriptions and phrases, and using other tricks of the writing trade to fool people into doing your bidding. Good business writers do not need to rely on such tricks. Instead they rely on effective, forceful writing to convey clearly what they mean to say to the reader. Figure 1.1 lists five easy tips for effective business writing.

Figure 1.1 Five Easy Tips for Effective Business Writing

1. Get to the point in the first sentence or paragraph.
2. Use simple, short words and sentences.
3. Make each point complete, clear, and logical.
4. Avoid unnecessary words.
5. Keep like things together and unlike things apart.

Types of Letters and Memos

Generally business correspondence comes in two forms: letters and memorandums. Letters are somewhat more formal and usually addressed to someone outside your business. Memos, on the other hand, are typically more informal and convey information to someone within your business. They are often shorter and convey only one or two pieces of information, while letters are often longer and expected to convey several pieces of information. So, for example, you may want to use a letter to sell an idea or a group of ideas to a customer or vendor and a memo to convey facts about a project to members of your building team.

We all receive form letters almost every day. Sophisticated business marketers have discovered that form letters are an inexpensive and powerful sales tool. With the advent of inexpensive personal computers and word-processing programs, even a small business can custom design personalized form letters.

Proper Format for Letters

The four main parts of any business letter are as follows:

1. return address and date
2. inside address and salutation
3. body of the letter
4. closing and signature

Businesses often choose one of several common styles for all correspondence. You can then arrange these four parts of any letter according to the company's style for correspondence. The choice of style is often one of personal preference, but letters will appear more professional if they all stick to one style.

If you use stationery that shows your company's address and telephone number, you only need to include the date rather than the return address and date. However, many businesses, particularly new firms, have not yet printed or cannot afford stationery. In those situations, include your company's full mailing address and telephone number at the beginning of each letter. The first few sample letters in Chapter 2 assume that the writer does not yet have preprinted stationery. After a few examples, the letters will omit the return address and assume that it appears on the stationery.

The inside address typically repeats the address that appears on the outside of the envelope. It is the address of the person to whom you are sending the letter. This is actually a functional part of the letter. In case you open a letter and throw away the envelope, you still have the inside address to help direct the letter to the proper person.

Also, the inside address should clearly spell out the title and, if appropriate, the department of the individual to whom it is sent. Many letters get misdirected within larger organizations, and this will help redirect the correspondence to the proper person. Sometimes you will send letters to a person whose name you do not know, for example, the director of marketing or credit manager. In these cases, the inside address is vital.

The salutation follows the inside address and usually takes the form of "Dear Mrs. Smith:" or "Dear Bill:" with a colon as the punctuation. You may abbreviate "Mr.," "Mrs.," "Ms.," and "Dr.," but you should spell out such titles as "Senator" or "Professor." Clearly, it is preferable to know the name of the addressee of your letter. This ensures that the person you intend to receive the letter actually receives it. This also saves time by preventing your mail from being rerouted. However, if you are unable to determine precisely who should receive your letter, then most authorities suggest using "Ladies and Gentlemen" or "Dear Sir or Madam."

The body of the letter follows the salutation. In the coming pages, this book will discuss in detail various types of body text for letters. Then, finally, the letter ends with a closing and a signature.

In days of old, closing phrases for business letters were lengthy, such as "Forever, I remain your respectful servant and confidant" or "With the utmost sincerity, I am your faithful companion." You have probably seen these endings to some of the formal correspondence in history books and museums. Today, a simple "sincerely," "cordially," or "very truly yours" will suffice. The meaning of these closings is the same as the longer, more formal closings. You are concluding the letter on a friendly note and wish to lend a formal yet warm good-bye to the end of the message.

In formal business letters, you type your full name and title below the closing and usually leave four blank lines for your signature. Most people have only semilegible signatures. If your company has even two or more people generating letters on the same letterhead, typing the full name and title of the person signing the letter will reduce confusion over who sent the letter. Adding your title after your name is also an important courtesy to the person receiving the letter. It can even add some force to the letter if your title happens to be president!

You will sign the letter after the closing and before your name and title. You may sign your full name if the letter is more formal or you do not know the person receiving the letter. Or you can sign your first name only if the letter is more informal and you do know the person receiving the letter.

Many letters take on the formality of legal documents. For example, a letter can state an intention to purchase a piece of property or a desire to commit to the sale of a truckload of lumber at a specified price. Remember that adding your signature to a letter carries with it a certain commitment, so you need to take seriously what you put into your business correspondence. If you think that a letter you are writing carries legal implications, you may also want to have a qualified attorney review it.

Perhaps the most common format for business correspondence is the blocked style. With this style, the date, return address, and signature all appear at the center of the page, while the other parts of the letter begin at the left margin. Letters are usually single spaced, paragraphs are not indented, and a blank line appears between paragraphs. In addition, three blank lines appear between the date and inside address. A single blank line appears between the inside address and the salutation and between the salutation and the first line of the letter. The semiblocked letter style is similar to the blocked style, but the first line of each paragraph is indented five spaces.

In the full-blocked style, all components of the letter line up along the left margin. Again, the paragraphs are not indented. The simplified style is similar to the full-blocked style, but a capitalized subject line replaces the salutation and the closing is omitted. The first few letters in Chapter 2 will show examples of all four of these formats for business letters.

The standard format for memos resembles the simplified style for letters. The return address, date, inside address, and salutation are replaced by "Date," "To," "From," and "Subject" lines. Memos also have no closing or signature.

Formal versus Informal Styles

Correspondence can have a range of formality, from handwritten postcards ("Gosh, France is great in the summer! Sure wish you could join us!") to very formal letters ("Unfortunately, Mr. Smith, the Primrose Plywood Company will no longer require your services as Controller...."). There are times for both of these extremes, but most of your business correspondence will fall somewhere in the middle.

In most business situations, you will want to send a formally typed letter—preferably on your company's stationery. However, you may have occasions where a more informal, handwritten note is appropriate. For example, a handwritten thank you note to your real estate agent for doing an excellent job of selling one of your homes adds a personal touch to your business correspondence. When that person is faced with selling another one of your homes, he or she may remember that personal note and work even harder to sell the next home.

As another example, you may find preprinted postcards more appropriate than form letters when surveying past and prospective customers about their tastes in home design. Therefore, carefully select the formality of your style according to the particular business situation you face.

Naturally, you wish to create a professional image for your company through your business letters. At the same time, you are in the selling business, so your correspondence with customers should carry an air of friendliness, as well as competency and correctness. Many building companies hire public relations consultants. An important aspect of their work should be an analysis of your correspondence style.

However, you can also analyze your correspondence on your own with four simple steps:

1. Read your letter out loud. Does it sound conversational? In other words, does it sound like what you would say to this person face-to-face?
2. If you were the person receiving this letter, what would you think of it? Would you understand it?

3. If you can afford the time, set the letter aside for a day and reread it. Does it still make sense?

4. When possible, have someone else read the letter and edit it. If no one else is available to read the letter for you, be sure to read your letter a second time out loud.

Polishing the Look of Your Letters

One of the first purchases of a new business is stationery. This can be a modest investment. Well under $100 should provide the start-up business with an ample supply of professionally printed letterhead and envelopes.

For many businesses, however, stationery can become a major production. If you hired public relations consultants, they would probably advise you to have the color, size, type of paper, and printing of your stationery custom designed to match your logo, signage, and other promotional material. In fact, the appearance of your stationery and your correspondence is an important part of your marketing strategy. A typical customer may receive 10 to 20 letters from you. The professionalism conveyed by this correspondence will go a long way to sealing a deal.

Additionally, stationery and correspondence that have a businesslike appearance can make a positive impression on lenders, suppliers, brokers, and other members of the business community. Picture a lender trying to assess your creditworthiness. The financial statements, business plan, and collateral look reasonable. However, if this information also has a businesslike appearance, wouldn't that further enhance your loan application? Would a lender want to risk a loan on someone who cannot even produce a presentable business letter?

What should go on your stationery? The answer may seem obvious—the company's complete name and mailing address, for instance. You would be surprised, though, at how often these are forgotten or incomplete on stationery. You should also include your company's complete telephone number because many letters stimulate return telephone calls. If you have a fax machine with a dedicated telephone number, you should include that number on your stationery as well. In addition, many com-

panies like to put their logos on their stationery for an added touch of professionalism.

Finally, if the street address of your business is different from the mailing address, then include the street address on the stationery. This will usually happen if the mailing address is a post office box. In fact, some vendors have a policy of not doing business with a firm that does not have a permanent physical location, so make sure that your street address does appear on your stationery. Many businesses also list addresses of different branch offices on their letterhead. Others list the names of subsidiary companies or separate divisions. For a summary of what you should put on your stationery, see Figure 1.2.

Another way to polish the look of your letters is to carefully check them for grammar, spelling, punctuation, and typographic errors. These four items can make a big statement about the professionalism of your company. Would you want someone building your house who is not careful enough to proofread business letters?

In addition, the fourth item—typographic errors—can seriously impact your business in other ways. For example, what if you bid on a remodeling job and sent the homeowner a letter saying, "We agree to perform the work for $8,000" when you really meant to say, "$9,000"? Or what if you said, "We guarantee to finish this work in 3 weeks"

Figure 1.2 What to Include on Your Company's Stationery

- Company's full name and logo
- Complete mailing address, including zip code
- Complete street address if different from mailing address
- Telephone number, including area code
- Fax number, including area code
- Division names if relevant
- Addresses for branch offices if relevant

when you meant to say, "13 weeks"? These errors happen every day and can be critical.

As mentioned earlier, try to have someone else read your letters before they go out the door. If that is not possible, take the time to reread each letter, preferably out loud, and carefully check for errors.

Business Correspondence Today And in the Future

Business correspondence normally refers to letters that are mailed in standard envelopes through the U.S. Postal Service. However, business correspondence has broadened greatly in the past few years to include methods of communication that previously only existed in science fiction.

With sophisticated word-processing programs available, you can produce most types of correspondence easier, faster, and cheaper than you could type them a few years ago. Now you can generate multiple personalized form letters to whole categories of potential customers. For example, a personalized letter announcing a new subdivision to every real estate agent in a particular neighborhood is now a simple reality. The better word-processing packages can even check spelling and print envelopes.

Desktop publishing has also become the rage. Many businesses use formal newsletters and brochures as vital marketing tools. You can now design and develop these pieces right in your own office with desktop-publishing software and a laser printer.

Many businesses now use fax machines, which allow rapid transmission of documents from one point to another for the price of a telephone call. You can typically transmit a one-page fax message in under a minute, and the long-distance charges for this anywhere in North America will be far less than comparable telegraph charges in the past.

Faxed documents range from scrawled notes to formally typed letters on stationery. Most companies use a cover form that accompanies all faxes. This cover form contains the following information about the fax:

- date and time of transmission
- name, title, and company of the person sending the fax
- name, title, and company of the person receiving the fax
- number of pages being transmitted, including the cover form
- telephone number of the machine sending the fax
- name and telephone number of the person to contact if problems arise with transmission

The fax cover form saves time because it tells at a glance exactly who should receive the fax, who sent it, and who to call if it fails to arrive.

Finally, many businesses are now adapting electronic mail for their communications needs, particularly between divisions of a company or among similar companies. With electronic mail computers literally talk to one another over the telephone lines, swapping messages without ever printing a hard copy. This new and growing field holds out the promise of becoming a practical tool for businesses in the near future.

The coming chapters will accomplish three things. First, they will offer a broad array of sample letters arranged in a logical order. The book begins with the basic letters a builder needs to start a business and progresses through letters for more complicated situations faced by an established builder. You are welcome to adapt the sample letters in this book for your own use. However, you will probably discover that you need to tailor them extensively to your own particular circumstances. In addition, always remember that, if you think a letter carries legal implications, you should have a qualified attorney review the letter.

Second, the book highlights some of the techniques used to structure a letter. Most of the ideas covered in this first chapter are explained with clear examples. After reading these, you should have a basic understanding of letter style, tone, and format.

Third and finally, hopefully you will garner an appreciation for the use of proper correspondence as an integral part of your business management and marketing strategies. Time and time again, business leaders are surveyed and asked to rank the skills necessary to succeed in their industry. Consistently, communications skills rank at the top of the list. Your ability to effectively communicate your ideas in writing to customers, subcontractors, suppliers, brokers, lenders, and others will help to determine the success of your business.

Chapter 2

Start-up Letters

Getting Started

The new builder who has just left a job to start a business needs to get the word out on the street as quickly as possible. To do this, the builder has two options. First, you can hire an advertising agency to put together an ad campaign. Second, you can begin by writing several business letters. In fact, even if you can afford the ad campaign, when you start a building business you need to write to several key people, including lenders, suppliers, brokers, and other members of the business community in your area. If you are resigning from a job to become a builder, the logical place to begin is with a letter of resignation as shown in Figure 2.1.

Figure 2.1 Letter of Resignation

```
                                    101 Hilldale Court
                                    Annandale, Virginia  20200
                                    January 24, 19XX


Mr. Jeffrey Smythe
President
Smythe Builders, Inc.
18 North Bay Bridge Road
Annandale, Virginia  20001

Dear Jeffrey:

     As we discussed today, I am submitting my resignation as
the construction superintendent of Smythe Builders, Inc.,
effective February 15, 19XX.

     After leaving Smythe Builders, I plan to establish Jason
Cartwright Builders, a homebuilding company that specializes in
upscale custom homes. Since Smythe Builders is primarily a tract
and commercial construction firm, our two companies should not
compete with one another.

     Thank you for your offer of help in this new venture. Any
guidance and support you can provide would be most welcome. In
addition, please call on me if I can assist you in any way. I
look forward to maintaining a close business relationship, as
well as a personal friendship, with you in the future.

                              Sincerely,

                              Jason Cartwright
```

Notice the letter in Figure 2.1 uses the semiblocked style. Also the letter highlights two important points. First, before Jason wrote this letter, he and his boss discussed the matter. This is a clear example of putting a conversation in writing soon after it happens to avoid confusion over what was said. Second, Jason and Jeffrey are reasonably good friends and hope to remain so. The positive tone of the letter, coupled with Jason's clear statements in writing that he wants to remain in close contact with Jeffrey, serves to reinforce this relationship. Unlike Figure 2.1, the letter in Figure 2.2 is written in the blocked style. Note the difference in paragraph indention.

Figure 2.2 Letter Requesting Leads from a Business Acquaintance

```
                                 101 Hilldale Court
                                 Annandale, Virginia  20200
                                 January 24, 19XX

Mr. Harold Osborne
Osborne Commercial Construction
415 Oak Street
East Annandale, VA  20002

Dear Harold:

You said I would never do it, but I did! Today I resigned from
Smythe Builders and plan to set up my own company next month. As
we discussed earlier, this company will specialize in building
upscale custom homes, particularly in the new neighborhoods down
by the bay.

On many occasions you have described for me the struggle you
went through when you started your business. Any additional
survival tips you might share with me now would be very helpful.
Also, since you specialize in paving the roads in the new sub-
divisions by the bay and our businesses will clearly complement
each other, would you happen to have any leads for me?

Thanks again for all of your encouragement, and I will call you
as soon as I am officially in business.

                         Sincerely,

                         Jason Cartwright
```

The letter in Figure 2.2 is professional and businesslike yet less formal than the letter in Figure 2.1. Note two important aspects of the letter in Figure 2.2. The first paragraph sets the tone for the letter and describes what the letter is about. That paragraph summarizes what is to come. It is short and to the point, so the busy reader can catch the heart of the message immediately. The last paragraph describes some action that will follow—a telephone call from Jason to Harold in the near future.

Another letter you may have to write when you start your building business is a letter to the contractor's licensing board as shown in Figure 2.3. Transferring a license from one company to another may seem like a simple task, but you will need to make many of these formal requests within the first few days of starting a new business.

Figure 2.3 Letter to the Contractor's Licensing Board

```
                                    Jason Cartwright Builders, Inc.
                                    101 Hilldale Court
                                    Annandale, Virginia  20200
                                    (703) 555-1492

February 15, 19XX

Contractor's Licensing Board
State Courthouse Building
Richmond, Virginia  22222

Dear Sir or Madam:

Effective today, I am establishing Jason Cartwright Builders. My
general contractor's license, No. 555-X22-GC, is currently list-
ed with Smythe Builders, Inc., of Annandale, Virginia, where I
was employed as the construction superintendent until today.

Please transfer my license to my new company at the above
address, effective immediately. Thank you.

                                    Sincerely,

                                    Jason Cartwright
```

A good business adviser, such as a certified public accountant or someone recommended by the Small Business Administration, can help you develop a list of agencies and departments that should receive initial start-up letters from you like the one shown in Figure 2.3. Many of these agencies have the authority to impose stiff penalties if you fail to contact them upon establishing a new business. Thus, as with many other steps you take in your business, putting these requests in writing and keeping a copy of them is a good way to establish a paper trail of your actions.

Another example of an initial start-up letter is an application to the municipal business licensing office as shown in Figure 2.4. While the letter to the contractor's licensing board was a simple request, the letter in Figure 2.4 explores a slightly more complicated matter.

Figure 2.4 Letter to the Municipal Licensing Office

```
                                    Jason Cartwright Builders, Inc.
                                    101 Hilldale Court
                                    Annandale, Virginia  20200
                                    (703) 555-1492

February 15, 19XX

Ms. Margaret McLeed
Municipal Licensing Board
City Office Building
Annandale, Virginia  20204

Dear Ms. McLeed:

A few weeks ago I established Jason Cartwright Builders, a
custom homebuilding company headquartered in my home at the
above address. While 101 Hilldale Court is not located within
```

the incorporated city limits of Annandale, I plan to build homes
in some of the new communities being developed within the city
limits.

At your earliest convenience, I would appreciate the opportunity
to discuss my municipal licensing needs with someone in your
office and will contact you to set up an appointment in the next
few days.

Thank you.

Sincerely,

Jason Cartwright

The letter in Figure 2.4 is an example of a full-blocked style of letter. As
Chapter 1 mentioned, most businesses will develop and use one format for all correspondence. Also, although Jason has yet to print stationery, he is typing his full
return address and telephone number centered at the top of the first page on all
letters to give them a more professional appearance. This is a perfectly acceptable
approach and a useful way of giving a businesslike appearance to your correspondence. From this point forward, the book assumes that Jason is using preprinted
stationery.

When you start a new building business, you might decide to hire a construction
superintendent. If you receive a referral on a prospective employee from a business
acquaintance, you should thank that person in writing as shown in Figure 2.5. Once
you have hired someone, to avoid confusion you need to put that person's job
responsibilities and terms of employment in writing as shown in Figure 2.6.

Figure 2.5 Letter Thanking a Business Associate for the Referral of an Employee

Jason Cartwright Builders, Inc.
101 Hilldale Court
Annandale, Virginia 20200
(703) 555-1492

February 27, 19XX

Mr. Harold Osborne
Osborne Commercial Construction
415 Oak Street
East Annandale, VA 20002

Dear Harold:

Thank you for referring Billy Joe Spears to me. His credentials
seem excellent, and I think he will make a good construction
superintendent.

As you know, I have one or two strong leads and will probably
begin my first custom building job in early April. However, a
superintendent needs to be on board well before then to coordinate subcontractors and suppliers. If Billy Joe is available,
I will probably want him to begin work around March 15.

Before Billy Joe begins work, can you advise me on a starting
salary for him? Although I have a few ideas on this matter, your

suggestions would be helpful. Later in the week I will call you
to discuss your suggestions. Thanks again for the referral.

Sincerely,

Jason Cartwright

Figure 2.6 Letter Offering a Job

Jason Cartwright Builders, Inc.
101 Hilldale Court
Annandale, Virginia 20200
(703) 555-1492

March 1, 19XX

Mr. Billy Joe Spears
14 Pine Street
Annandale, Virginia 20005

OFFER OF EMPLOYMENT

This letter confirms our conversation this morning and outlines
the details of the offer of employment we discussed. Effective
March 15, 19XX, Jason Cartwright Builders, Inc., will employ you
as a construction superintendent. Your monthly rate of pay will
be $_____.__, and it will be paid to you every two weeks. In
addition, you will be paid ___% of any amount under budget on
any job under your supervision.

You will be allowed one week of paid vacation annually after 12
months of continuous employment and two weeks of paid vacation after three years of employment. This vacation will not
accumulate. You will accrue one day of sick leave for each two
months of employment, and this may accrue to a total of 30 days.
You may also elect to be covered under the _____
health plan.

At Jason Cartwright Builders you will have the following duties:

1. You will handle all phases of construction on any job under
your supervision.

2. You will supervise and schedule all subcontractors and any
tradespeople and laborers directly employed by the company. You
will use contract forms and procedures provided by the company
to engage these subcontractors.

3. You will be responsible for all materials ordered to a job
and will use the purchase order system provided by the company
to order those materials.

4. For each job under your supervision, you will develop budgets
and schedules, working with me and any salespeople involved,
before the acceptance of the contract on the job.

The company will provide you with a company truck, a gas credit
card, and a telephone. If you accept the terms outlined in this
letter, please sign the two enclosed copies of the letter,
return one signed copy to me, and retain one signed copy for

your records. Please note that this letter is not an employment contract.

Once you have returned one signed copy of this letter to me, please call me within the next week so we can make the final arrangements to bring you on board.

Thank you, and I look forward to working with you.

Jason Cartwright (Date)	Billy Joe Spears (Date)
President	

Enclosures

The simplified style used in the letter in Figure 2.6—modified with the two blank lines at the end for signatures—is appropriate for this type of correspondence. This letter shows one example of how to put a job offer in writing. Figure 3.7 will show another example of how an established company might put a job offer in writing.

Informing the Business Community of the New Venture

As soon as your business opens its doors, your marketing campaign should begin. The local home builders association (HBA) is a natural place to start this campaign. Keeping the association's executive officer informed of your new business venture is a good way to begin to establish a reputation for professionalism in your community. Figure 2.7 shows a sample letter to an HBA executive.

Figure 2.7 Cover Letter for an Application to an HBA

Jason Cartwright Builders, Inc.
101 Hilldale Court
Annandale, Virginia 20200
(703) 555-1492

March 1, 19XX

Mr. Zebediah Long
Executive Director
Home Builders Association
  of Greater Annandale
Post Office Box 62
Annandale, Virginia 20007

Dear Mr. Long:

Enclosed is my application for membership in the Home Builders Association of Greater Annandale. My former employer of 12 years and good friend, Jeffrey Smythe of Smythe Builders, Inc., is sponsoring me. Mr. Smythe has actively participated in the local association for many years and has always spoken highly of the benefits of his membership. In the years to come I also hope to become an active member of the association.

My new company, Jason Cartwright Builders, Inc., will specialize in upscale custom homes in the new communities developing on the bay. I plan to build five to seven homes each year, although my

```
business plan does not anticipate reaching this level for two
years.

Next month I intend to print a marketing brochure that will
outline my experience and plans for the future. I will send you
a copy of the brochure as soon as it is finished. In the mean-
time, if any questions arise about Jason Cartwright Builders,
please do not hesitate to call me at your convenience.

Sincerely,

Jason Cartwright
```

The letters in Figures 2.8 through 2.10 follow a marketing technique often called the up-down-up principle. Most well-established salespeople, loan officers, and purchasing agents are often too busy to give a new-customer salesperson very much time. Think for a moment about real estate agents. As a builder, do you want an agent who is on the go 18 hours a day to represent your homes? Or do you want the person who sits in the office waiting for the telephone to ring to represent your homes?

As a new builder, you need to show your product to sales agents who will in turn become your marketing representatives to the general public. The challenge, of course, is getting the attention of the on-the-go agents who will make things happen for you. The up-down-up principle helps you get the attention of the people you need to see. The principle works this way. First, get the attention of someone up the ladder, such as the president, broker-in-charge, or office manager. Let the manager briefly know exactly what you want. Then the manager will do the rest. This principle may sound too simple, but it works most of the time. In fact, many large marketing firms teach this as a science to new sales employees.

Figure 2.8 Letter to a Prospective Supplier

Jason Cartwright Builders, Inc.
101 Hilldale Court
Annandale, Virginia 20200
(703) 555-1492

```
March 1, 19XX

Mr. Jon Fox
President
Fox Specialty Products Company
85 North Bay Bridge Road
Annandale, Virginia  20008

Dear Mr. Fox:

Your firm comes highly recommended as an excellent source of
hardware and other specialty items for finer homes. Recently,
after 12 years with Smythe Builders, I have formed my own
company to specialize in building a limited number of upscale
homes for discriminating buyers.

Given the size and complexity of the hardware requirements for
each home we plan to build, Jason Cartwright Builders needs to
work with a knowledgeable and experienced representative of your
```

company. Would you have an appropriate person from your staff call on me to discuss these needs further?

Thank you, and I look forward to hearing from you soon.

Sincerely,

Jason Cartwright

Figure 2.9 Letter to a Prospective Lender

Jason Cartwright Builders, Inc.
101 Hilldale Court
Annandale, Virginia 20200
(703) 555-1492

March 1, 19XX

Mr. Benjamin Douglas
President
Commerce Bank of Annandale
Annandale, Virginia 20009

Dear Mr. Douglas:

After 12 years with Smythe Builders, Inc., most recently as the construction superintendent, I have formed my own company to specialize in building a limited number of upscale homes for discriminating buyers.

Fortunately, I have been planning the establishment of Jason Cartwright Builders for over two years. As you can see from the enclosed business plan, the company will be more than adequately capitalized.

The business plan also uses reasonably conservative assumptions to determine the cash-flow requirements for the company during its early stages of growth. However, we will need a line of construction credit and equipment financing and would benefit from keeping this financing in one place.

On several occasions, Jeffrey Smythe has described to me his long and profitable relationship with your institution. On his recommendation, I would like to discuss my borrowing needs with one of your loan officers. Would you please have the appropriate person call on me within the next week?

Thank you, and I look forward to hearing from your institution soon.

Sincerely,

Jason Cartwright

Enclosure

After reading the letter in Figure 2.9, Mr. Douglas will probably scribble a note in the margin and route the letter to one of the bank's loan officers. This up-down-up approach to business communications does two things. First, the people up the ladder, who will ultimately decide on your line of credit, will have heard your name before. Furthermore, Mr. Douglas's note will get the attention of one of the better loan officers—the very type of person you want to have working on your account. The up-down-up principle also works in other fields. In Figure 2.10 notice how this approach works with a real estate broker.

Figure 2.10 Letter to a Prospective Broker

Jason Cartwright Builders, Inc.
101 Hilldale Court
Annandale, Virginia 20200
(703) 555-1492

March 1, 19XX

Ms. Eva Forsythe
Broker-in-Charge
Forsythe Realty
1919 Junction Way
Annandale, Virginia   20010

Dear Ms. Forsythe:

After 12 years in the homebuilding industry, I have established Jason Cartwright Builders, which will specialize in one-of-a-kind custom homes for discriminating buyers. We plan to build approximately six homes each year and intend to focus on the new neighborhoods on the bay.

As you know, your agency has an excellent reputation for representing discerning clients. I think that Jason Cartwright Builders will offer homes that your clients will want to see. You currently work with some of the better builders in the bay area, and we would be honored if our company could join that roster.

Several of your top people were named to the Million Dollar Club this year. I would appreciate the opportunity to meet with one of them and discuss what we have to offer. Please have one of them call on me in the next week or so.

Thank you, and I look forward to hearing from your company soon.

Sincerely,

Jason Cartwright

In Figures 2.8 through 2.10, the first paragraph of each letter sets a positive, action-oriented tone for the rest of the letter. Furthermore, the second-to-last paragraph asks the person receiving the letter to take some action in the near future. However, remember that in these letters you are asking a total stranger to do something for you. In Figure 2.8, ultimately you want the supplier to open an account and extend you a line of credit. You do not say that explicitly, but that is the implied message. In Figures 2.9 and 2.10 you are more open with the banker and broker. You want the former to lend you money, and you want the latter to add your homes to her product line. These are tall requests, so your letters to these people need to be brief, polite, and subtle sales pitches.

The letter in Figure 2.11 introduces Jason Cartwright to a subdivision developer. As a new builder, Jason will need to become involved in many new projects and will want to be included in the earliest planning stages of new communities.

Figure 2.11 Letter to a Subdivision Developer

Jason Cartwright Builders, Inc.
101 Hilldale Court
Annandale, Virginia 20200
(703) 555-1492

```
March 1, 19XX

Ms. Mary Lewis
Lewis and Clark Development Company
#1 Missouri Way
Annandale, Virginia  20011

Dear Ms. Lewis:

Your two newest communities, New Louisiana and Northwest Pas-
sage, are very impressive. As a result of your rigorous archi-
tectural review process, you are preserving the continuity and
value of these fine neighborhoods.

In February I established Jason Cartwright Builders, a company
that will specialize in building one-of-a-kind homes for dis-
criminating buyers. New Louisiana and Northwest Passage are
exactly the types of communities where we want to build, and
your clients are precisely the type of people we want to serve.

At your convenience, I would appreciate the opportunity to meet
with you and any appropriate members of your staff to discuss
what we have to offer. Within the week I will call you to set up
an appointment. Until then, if you have any immediate questions,
please do not hesitate to call.

Sincerely,

Jason Cartwright
```

So within the first few weeks of setting up his new company, Jason has contacted several key people and companies by letter. Naturally he will develop an ongoing relationship with these people as his business grows and prospers. The discussion will return to the situations faced by a custom builder like Jason a little later in the chapter. For now, the discussion turns to some of the special correspondence used by a speculative builder.

Building a Speculative Home

Building a new home, whether on speculation or contract, will tend to disrupt the lives of the existing homeowners. Of course, you can never please everyone, but you might change potentially irritable neighbors into salespeople by dropping each of them a friendly note as shown in Figure 2.12.

Figure 2.12 Form Letter to Neighbors

Sanderson, Inc.
Post Office Box 82
8818 Orange Avenue
Kennebunkport, Maine 97531
(202) 456-6797

```
March 1, 19XX

Mr. and Mrs. George Robertson
1422 Oceanside Parkway
Kennebunkport, Maine  97012

Dear Mr. and Mrs. Robertson:

Please let me take this opportunity to introduce myself. I am
John Blakeley, president of Sanderson, Inc., one of Maine's
oldest and best-respected homebuilding companies. My grand-
father, Ezra Sanderson, founded this company in 1912 and in his
lifetime built many fine homes all across Maine.

Within the next month Sanderson will begin building a home in
your neighborhood at 1509 Oceanside Parkway. The city government
has issued all of the necessary permits, and next week we will
finish the engineering studies.

You and your neighbors will be curious as the construction
progresses, but for the sake of safety we ask that everyone,
particularly children, stay away from the construction site. As
soon as the home is ready to show to the public, Sanderson will
hold an open house and invite all of the neighbors.

In our experience, the most satisfied homebuyers are those who
move into a neighborhood where they already have friends and
acquaintances. If you know anyone who might want to buy this
home and move into your community, you might want to mention the
special early-bird discounts and options available to them
during the first few weeks of construction. If so, please give
us a call and we will gladly give you more details.

Sanderson Builders is pleased to make this new contribution to
your community. If you have any questions, please give us a call.

Sincerely,

John Blakeley
```

Word processors are wonderful inventions! With most of the newer programs, you can customize the form letter in Figure 2.12 and send it to everyone living on Oceanside Parkway. The program can even type the envelopes for you. With little effort a form letter can become a powerful public relations and marketing tool.

As a new speculative builder, you must take care of many other nitty-gritty tasks when you start to build a home. The letters in Figures 2.13 through 2.16 are just a few examples of several kinds of letters you might need to write as you build the house.

Figure 2.13 Letter Requesting a Bid from a Subcontractor

Sanderson, Inc.
Post Office Box 82
8818 Orange Avenue
Kennebunkport, Maine 97531
(202) 456-6797

```
March 2, 19XX

Mr. Jerry River
Jerry River's Septic Service
808 Buried Treasure Boulevard
Kennebunkport, Maine   97013

Dear Jerry:

This month Sanderson Company will begin construction on a new
single-family home at 1509 Oceanside Parkway. On the recommenda-
tion of my good friend, Hal Batson, we would like to have you
submit a bid for providing the septic system for this home.
Enclosed are the plans and specifications for your review.

Please include the following items as part of your bid:

1.  names of three other contractors or homeowners you have done
work for in the past year

2.  copies of all relevant licenses

3.  copies of your workers' compensation and public liability
insurance policies

4.  your signature on the specifications, indicating that you
have read and understood them

Sanderson will have a site meeting on March 12, and we would
like to have you join us. If this date is inconvenient for you,
please call and check with our superintendent, Frank Holliday,
about an alternate date. Given our current schedule, you need to
submit your bid along with all other documents by close of
business on March 15.

Thanks, Jerry. I look forward to hearing from you.

Sincerely,

John Blakeley
```

A request for a bid as shown in Figure 2.13 may be handled more or less formally in different parts of the country. Established builders in some parts of the country may use formal bid request procedures. In other parts of the country, this process is much less formal. Once you have opened the bids and accepted the winning bid, you should send a letter of confirmation, outlining the remaining procedures.

Figure 2.14 Confirmation Letter to a Subcontractor

Sanderson, Inc.
Post Office Box 82
8818 Orange Avenue
Kennebunkport, Maine 97531
(202) 456-6797

March 17, 19XX

Mr. Jerry River
Jerry River's Septic Service
808 Buried Treasure Boulevard
Kennebunkport, Maine 97013

Dear Jerry:

Congratulations on submitting the winning bid to provide the septic system for the home that Sanderson will build at 1509 Oceanside Parkway. We look forward to having your company work on this project.

Our office has received copies of your current licenses and insurance papers. Please remember that you are responsible for ensuring that we have the most current version of these licenses and insurance papers on file and to notify us immediately of any changes.

Enclosed is a tentative schedule for this job. The superintendent will contact you by telephone and letter to confirm this schedule about 30 days before you actually begin work.

Also enclosed are two copies of a purchase order for this work. The purchase order outlines methods and amounts of payments, conditions for this work, and other policies. By signing and returning this purchase order, you agree to accept these procedures and policies as conditions of the work.

In addition, the purchase order stresses two points that are very important to Sanderson, Inc.

First, safety is key with us. If your people are observed working in an unsafe manner, this is cause for immediate termination of the contract and forfeiture of retainage. We regret being harsh on this point, but safety is critical in the building business.

Second, keeping the job site neat and attractive is extremely important. Although construction is a messy business, it is also a selling business. We must all remember that we are putting on a show for the customer, who pays our salaries.

We at Sanderson look forward to working with you in the coming months. Please give me a call if you have any questions.

Sincerely,

John Blakeley

Enclosures

Notice how the letter in Figure 2.14 puts important points about safety and tidiness into separate paragraphs for emphasis. This is an effective technique to use in your letters when you want to emphasize or clarify a point.

As you know, occasionally the subcontractor's work does not go as planned and you need to send a letter encouraging the subcontractor to finish the job as shown in Figure 2.15. Note the use of the simplified style for emphasis in the letter in Figure 2.15. Using the simplified style adds some forcefulness to the letter, which is just what is needed in this situation.

Figure 2.15 Letter Encouraging a Subcontractor to Finish a Job

Sanderson, Inc.
Post Office Box 82
8818 Orange Avenue
Kennebunkport, Maine 97531
(202) 456-6797

July 30, 19XX

Mr. Jerry River
Jerry River's Septic Service
808 Buried Treasure Boulevard
Kennebunkport, Maine   97013

SUBJECT: DELAYS IN SUBCONTRACTED WORK

On March 17, 19XX, your company was awarded the contract to provide the septic system for the house at 1509 Oceanside Parkway. On April 1, the construction superintendent notified you both verbally and in writing to begin work on May 1.

Your original subcontract bid (see attachment) specified a completion time of no more than 30 days after such notification, barring such unexpected hazards as the document specifies. This completion time was then incorporated into the purchase order dated March 17, and you were reminded of this in your notification of April 1.

However, your company did not begin work at 1509 Oceanside Parkway until May 15. This delay by itself causes serious delays to other subcontractors and the job as a whole. However, you have also not yet finished the job, 76 days after beginning the work and 91 days after you were notified to begin the work.

Further, the construction superintendent has determined that the job is not yet half finished, which implies at least another 15 days of work even if you can complete this in a timely fashion.

The construction superintendent discussed this matter with you in person on June 1 and 10 and July 7, and by telephone on July 2 and 15. From all appearances, no unforeseen hazards have arisen, and you have not mentioned any unforeseen hazards.

At this point, you must finish the job within 15 days. If your company does not begin work tomorrow and make timely progress toward finishing the job within 15 days, we will then consider this contract terminated and any retainage forfeited as specified under the terms of the contract.

Jerry, your company came highly recommended to us for this
particular job, and once you have met your contractual obligations, hopefully we can put this matter behind us. If
Sanderson can assist you in preventing further delay on this
job, please let us know. In addition, if you want, I would be
happy to discuss this job with you on the telephone or in
person.

John Blakeley

Notice that John Blakeley carefully outlines the problem in the first part of the letter in Figure 2.15 by using factual statements. Obviously, the construction superintendent of this company also had to keep good records of any conversations with Jerry River's Septic Service, so that person could back up verbal requests with follow-up letters.

In the body of the letter, John makes carefully worded demands. He does not try to play lawyer. He simply and forcefully explains the contract requirements and their rationale, the exact result if these requirements are not met, and the justification for those results. In the final paragraph of the letter, John holds out some hope that the relationship can be salvaged. This is important and concludes an otherwise tough letter on as friendly a note as possible under the circumstances. Again, if you are writing a letter that might involve a contract dispute, you may want to have it reviewed by a qualified attorney.

Just as you may need to send a letter encouraging the completion of a job to a subcontractor, you may also occasionally need to send a similar kind of letter to a supplier as shown in Figure 2.16. Even regular suppliers, who are used to getting nothing but purchase orders and checks, need some reminders and encouragement now and then.

Figure 2.16 Letter to a Senior Staff Member at a Supplier

Sanderson, Inc.
Post Office Box 82
8818 Orange Avenue
Kennebunkport, Maine 97531
(202) 456-6797

April 21, 19XX

Ms. Diana Buckingham
Buckingham Hardware
#2 Cloisters Lane
Kennebunkport, Maine   97016

Dear Diana:

Your company and mine have been doing business together for over
eight years. The level of service we receive from your company
is excellent. However, yesterday one of your trucks delivering
hardware to 1509 Oceanside Parkway destroyed three trees that we
had carefully marked to save for future landscaping.

Although the construction superintendent was at lunch when this
happened, we are quite sure the damage was done by one of your
trucks because hardware was the only thing scheduled for deliv-
ery yesterday. Also, when the driver hit the largest of the

```
three trees, the passenger-side door fell off, and it has your
company's name and crown-shaped logo on it.

Enclosed is an estimate from the landscaping company for the
repair work. Please send us a check for that amount or deduct it
from our account. In addition, we still have the truck door and
can return it to you if you wish.

Thank you for your help on this matter. Give my best to Chuck
and the kids, and I hope we see you next month at the home
builders annual picnic.

Sincerely,

John Blakeley
```

See? A complaint letter does not have to be nasty. However, it does have to clearly and precisely discuss the issue at hand.

Selling a Speculative Home

Up to this point, only the letter in Figure 2.12 has dealt even slightly with the marketing and sale of a speculative home. Not surprisingly, sales-oriented letters can provide a great benefit to your company at a low cost. As shown in Figure 2.17, one of the most important sales letters you will ever write is a personalized form letter announcing an open house.

Figure 2.17 Letter Announcing an Open House

Sanderson, Inc.
Post Office Box 82
8818 Orange Avenue
Kennebunkport, Maine 97531
(202) 456-6797

```
August 1, 19XX

Mr. and Mrs. George Robertson
1422 Oceanside Parkway
Kennebunkport, Maine  97012

Dear Mr. and Mrs. Robertson:

Sanderson, Inc., cordially invites you to the long-awaited open
house for 1509 Oceanside Parkway on Saturday, September 1.

Several of the leading real estate agencies from Kennebunkport
will join us with prospective buyers, so this is an opportunity
to come and meet your potential new neighbors.

The open house will begin at 9:00 a.m. and continue until
sundown. We will have refreshments on hand and even
entertainment for the kids.
```

Thank you for your patience with our construction site, and we look forward to seeing your family at our open house.

Sincerely,

John Blakeley

Once you hold an open house, you need to send a follow-up personalized form letter to prospective customers as shown in Figure 2.18. Then, once you have a customer under contract, you may also need to correspond with that person in writing. Figure 2.19 shows an example of such correspondence.

Figure 2.18 Personalized Form Letter to Prospects

Sanderson, Inc.
Post Office Box 82
8818 Orange Avenue
Kennebunkport, Maine 97531
(202) 456-6797

September 3, 19XX

Mr. and Mrs. Hiram Henry
8818 Shreveport Highway
Knoxville, Tennessee 11218

Dear Mr. and Mrs. Henry:

Thank you for joining us at the open house last Saturday to meet some of the neighbors from Oceanside Parkway. We are delighted to hear that you are planning to buy a home in Kennebunkport. Many fine families either vacation here or enjoy it year-round.

As the Trapp Agency probably mentioned, Sanderson has built homes all over New England over the past 80 years. Quality is our number one priority, and we consistently receive top ranking in customer satisfaction. Working with your agents, we would be happy to provide you with any information you need to make this important decision.

Thank you again for joining us on Saturday, and I look forward to seeing you in the near future.

Sincerely,

John Blakeley

Figure 2.19 Letter to Customers under Contract

Sanderson, Inc.
Post Office Box 82
8818 Orange Avenue
Kennebunkport, Maine 97531
(202) 456-6797

November 17, 19XX

Mr. and Mrs. Hiram Henry
8818 Shreveport Highway
Knoxville, Tennessee 11218

Dear Mr. and Mrs. Henry:

Construction on your new home is moving along, and as we
discussed last month, we are still planning for a January 31,
19X1, closing.

Enclosed is a copy of your color selections sheet. Please review
the final color selections for your home and notify us in writing within the next 10 days if any of these colors need to be
changed.

As you know, your contract calls for a $50 charge per change
order at this point in construction. However, you were very
patient when we ran into delays with the improper septic system
last month. To express our appreciation, we waive any change
order charges, as long as you have those orders to us by the end
of next week.

Working for you has been a pleasure, and we at Sanderson wish
you many happy years in your new home.

Sincerely,

John Blakeley

As people who have worked in the building business for awhile know, home
financing runs in cycles. One day, every permanent mortgage lender in town is
looking for business. The next day, or so it seems, the same lenders only open their
offices from 11:00 to 11:15 a.m. Once you have a customer under contract, you may
have several occasions to write letters to lenders. Figures 2.20 and 2.21 show
sample letters to a permanent lender and a construction lender under those circumstances.

Figure 2.20 Letter to a Permanent Lender

Sanderson, Inc.
Post Office Box 82
8818 Orange Avenue
Kennebunkport, Maine 97531
(202) 456-6797

September 15, 19XX

Ms. Sharon Marcos
Vice President
Kennebunkport Bank of Commerce
Post Office Box 7
Kennebunkport, Maine 97520

Dear Ms. Marcos:

As you know, Mr. and Mrs. Hiram Henry have selected the Kennebunkport Bank of Commerce (KBC) as their mortgage lender. This
is the first time that one of our clients has dealt with KBC,
and we hope that it is the beginning of a long and mutually
beneficial relationship.

Enclosed are three copies of the plans and specifications for
the Henrys' home. We are looking forward to a December 15
closing, barring any unexpected circumstances. If this date for
closing changes at our end, we will let you know as soon as
possible and we would appreciate the same notice from KBC.

To maintain a smooth closing process, Sanderson follows a few
simple steps to coordinate the process with the mortgage lender.
Janet Jameson is the person at Sanderson who tracks all loans
and closings for the customer. Before a closing, Janet contacts
the lender once a week to check on the status of the loan in
progress.

To do this for the Henrys, Janet will need the name of the
contact person in your office. She will then call that person in
the next few days to discuss the Henrys' loan. If your under-
writers require any other documentation from us for this loan,
please let her know at that time.

About two weeks before closing, Janet will need a detailed list
from the closing agent of all documents that Sanderson is required to provide at closing. Even when the closing is routine,
having the closing agent sign off on this list is very helpful.
Please let us know if you need any further information from us
as you work with the closing agent to develop this list.

Thank you for your help on this project, and we at Sanderson
hope to have many other opportunities to work with you and KBC
again in the future.

Sincerely,

John Blakeley

Enclosures

Figure 2.21   Letter to a Construction Lender

Sanderson, Inc.
Post Office Box 82
8818 Orange Avenue
Kennebunkport, Maine 97531
(202) 456-6797

```
September 15, 19XX

Mr. Fred Williamson
Vice President
Slade Savings Bank
44 Rock Harbor Way
Bedrock, Maine  97321

Dear Fred:

As we discussed by telephone, Sanderson, Inc., now has a signed
contract on the home that Slade Savings Bank is financing at
1509 Oceanside Parkway. As stipulated by our agreement dated
May 15, 19XX, we would like this home removed from the specula-
tive list, which will free up the face amount of the construc-
tion loan for further construction.

At your convenience, I would also like to meet with you to
discuss our upcoming line-of-credit renewal. As you know, given
Sanderson's excellent track record, this renewal has been rela-
tively routine in the past two years. However, our business plan
for the coming year calls for some expansion in our speculative
building, and we would like to count on Slade Savings to finance
a portion of that expansion.

In the next few days, let's schedule a meeting to discuss this
plan before we submit a formal proposal. As always, it is a
pleasure doing business with you and Slade Savings Bank.

Sincerely,

John Blakeley
```

The letter in Figure 2.21 kills two birds with one stone. First, you should always keep construction lenders informed of the status of any homes they finance. This is more than just good manners, it is good business. Of course, the sale of a house is a piece of good news, and you can use this good news to maximum advantage. In this case, John wants to set the stage for an increase in his construction line of credit and he uses the good news of a sale as an opportunity to do so.

Just before closing, you will want to send a letter to your customers preparing them for the closing process as shown in Figure 2.22. Among other things, this letter will describe your walk-through and warranty program. As you prepare this letter, keep in mind that many customers face the final steps of buying a home with a certain amount of nervousness. To help calm their fears, make your letter as simple, clear, and direct as possible.

Figure 2.22 Preclosing Letter to a Customer

Sanderson, Inc.
Post Office Box 82
8818 Orange Avenue
Kennebunkport, Maine 97531
(202) 456-6797

January 20, 19XX

Mr. and Mrs. Hiram Henry
8818 Shreveport Highway
Knoxville, Tennessee 11218

Dear Mr. and Mrs. Henry:

Thank you for buying a Sanderson home. On January 31, we are scheduled to close on your new home. This letter describes how Sanderson will assist you during the closing process.

First, you have either recently received or will soon receive a detailed list of instructions from the closing agent and mortgage lender. Our office has received a similar list in our capacity as the seller. If you have any questions about the closing process, please route them through your real estate agent to the mortgage lender.

However, we will work directly with you on a few other matters. First, Sanderson is a long-standing member of the Northeastern Warranty Program (NWP). This is a broad-based program that provides 10 years of structural limited warranty on your home. At closing we will provide you with the necessary insurance papers and other documents on this program.

In addition, Sanderson has an in-house customer service program that goes beyond NWP. We are proud of our commitment to service, and over the years our customers have been uniformly pleased with this program. As a first step, we will ask you to meet with us two days before closing to walk through your new home. Next week we will call to schedule a convenient time for this walk-through.

During the walk-through we will discuss every component of your new home in detail and go point by point over an 11-page checklist, often called a punchlist, to ensure that every item in your home meets or exceeds your expectations. Then we will both sign the checklist. If anything is noted on the checklist, our construction superintendent will try to follow up on each agreed-upon item within 30 days.

Again, if you have any questions about closing, please route them through your real estate agent to the lender or closing agent. If you have any questions about your new home, please give us a call. Thank you again for buying a Sanderson Home, and we look forward to seeing you on January 29 for your walk-through.

Sincerely,

John Blakeley

The first night in a new home can be one of the scariest experiences ever for your customers. Most of their furnishings are still packed in boxes or scattered around on the floor. Their new house has strange smells and sounds they have never experienced before. Most importantly, though, you now have their money and have gone away.

Many builders take this moment as an opportunity to reinforce their marketing efforts. Some builders have a surprise package dropped off at the new home while the buyers are at closing. This package can include some moving-in necessities, such as a roll each of paper towels and bathroom tissue, a bar of soap, or perhaps some air freshener. A family with several children would appreciate a gift certificate for a free lunch or dinner at a local restaurant. As shown in Figure 2.23, you can also reinforce your marketing efforts with a follow-up letter to the customer shortly after closing.

Figure 2.23 Postclosing Letter to a Customer

Sanderson, Inc.
Post Office Box 82
8818 Orange Avenue
Kennebunkport, Maine 97531
(202) 456-6797

February 5, 19XX

Mr. and Mrs. Hiram Henry
1509 Oceanside Parkway
Kennebunkport, Maine 97012

Dear Mr. and Mrs. Henry:

How are you settling into your new home? We at Sanderson hope
your move was pleasant and you are now enjoying your new home.
At closing you received a lot of information, so we want to
remind you of a few key items.

First, our construction superintendent tells me that work on the
punchlist items for your home is currently on schedule. If the
schedule changes in any way, we will let you know.

You received specimen copies of your NWP warranty papers at
closing. The actual insurance documents should be mailed to you
within six weeks.

Several of the key mechanical components of your home, such as
the hot water heater, kitchen appliances, and heating system,
have separate warranties and optional extended maintenance
contracts. These documents were given to you at closing. Please
be sure to review them at your earliest convenience.

Finally, the county tax collector will mail the property tax
bill, which is due at the end of the year, to our office. At
closing, you were given a credit for one month's worth of property taxes. At the end of the year, we will forward this tax
bill to you, and you should pay it at the county tax collector's
office. If you have any questions about this bill, please
contact your closing agent.

If you have any other questions about your new home, please give us a call. Thank you again for buying a Sanderson home.

Sincerely,

John Blakeley

Now, wasn't the postclosing letter in Figure 2.23 a nice thing to do? A successful clothing store owner once said, "I can train a chimp to put shirts in a bag and operate the cash register, but I can't get him to say, 'Thank you for your business!'" A pleasant postclosing letter that thanks customers for their business is one way to ensure referrals from past customers. It makes good marketing *and* good business sense. While you are at it, you also need to thank your listing and selling agents as shown in Figures 2.24 and 2.25.

Figure 2.24 Postclosing Letter to a Listing Agent

Sanderson, Inc.
Post Office Box 82
8818 Orange Avenue
Kennebunkport, Maine 97531
(202) 456-6797

February 5, 19XX

Ms. Julie Nessen
Senior Broker
Clearwater Realty
Post Office Box N
Kennebunkport, Maine 97010

Dear Julie:

Thank you for the excellent job your company did in marketing the home at 1509 Oceanside Parkway. I was really impressed with how quickly we had a contract and hope to work with you and Clearwater Realty again in the future.

Thanks to your hard-working staff, the open house at 1509 Oceanside Parkway was a great success. Most importantly, that is where we met the Henrys, who ultimately bought the home. In the future, however, we may want to consider inviting a few more agents from a broader array of firms.

You and your staff handled the delay in construction beautifully. That was our responsibility, and we will try to avoid having that happen again. If it does, though, let's you and I try to maintain closer communications with the buyers, so together we can avoid surprising them like we did this time.

Again, thank you for a job well done. When you have a chance, let's plan to meet and discuss another job with Sanderson.

Sincerely,

John Blakeley

cc: Sidney Barton, Broker-in-Charge

Figure 2.24 shows the first use of the "cc:" notation. It literally stands for "carbon copy." Even though carbon paper is not used any more for making copies of letters, that notation identifies the people who will receive copies of the letter. In this case, John is letting Julie know that her boss, Sidney, will get a copy of the letter. Sending a copy of a letter of thanks or praise to the supervisor of the person receiving the letter is generally a good practice.

Figure 2.25 Postclosing Letter to a Selling Agent

Sanderson, Inc.
Post Office Box 82
8818 Orange Avenue
Kennebunkport, Maine 97531
(202) 456-6797

```
February 5, 19XX

Ms. Helga von Trapp
The Trapp Agency
1819 Austria Way
Kennebunkport, Maine 97010

Dear Ms. von Trapp:

Thank you for putting together the Henrys' contract. We under-
stand that you worked with the Henrys for several months to
find them just the right home. Thank you again for placing your
confidence in Sanderson.

In your area we have two other homes for sale. One is a moder-
ately priced, move-up home, which is perfect for a growing fam-
ily. The other is an executive home with all the features one
expects in such a distinctive residence.

At your convenience next week, I would like to meet and discuss
with you the marketing of these two homes. You provided us with
such helpful insights on the Henrys' contract that you are sure
to have some fresh marketing ideas and strategies for these
other two homes.

In the next few days, let's schedule a meeting for next week,
and thanks again for your efforts on the Henrys' contract.

Sincerely,

John Blakeley
```

The second and third paragraphs of the letter in Figure 2.25 raise another issue. Thank you letters usually put people in a good mood, so they offer you an excellent opportunity to bring up other subjects. Of course, you can just say thank you and be done with it, but you would waste a great opportunity to do business.

Building and Selling a Custom Home

Many of the letters used in building a custom home are similar to those used in building a speculative home. Figures 2.26 through 2.30 show some samples of letters unique to the building of custom homes. One of the most important letters you can write to a custom home client is the precontract letter as shown in Figure 2.26.

Figure 2.26 Precontract Letter to a Client

Jason Cartwright Builders, Inc.
101 Hilldale Court
Annandale, Virginia 20200
(703) 555-1492

June 18, 19XX

Mr. and Mrs. Winfred Scott
123 Village Parkway
Annandale, Virginia 20026

Dear Mr. and Mrs. Scott:

Thank you for the opportunity to meet with you yesterday evening and discuss plans for your new home. As we discussed, we will have our proposal ready and delivered to you on June 26.

Having a home built is a significant undertaking. In order for you to feel comfortable hiring Jason Cartwright Builders, I have enclosed three letters of reference. One of the letters is from my former employer, Jeffrey Smythe, and tells you more about my ability to build your new home.

Also enclosed is a list of suppliers and subcontractors with whom I normally do business. Please feel free to contact any of these people for references as well.

Again, thank you for the opportunity to submit a proposal on your new home. I look forward to becoming your builder.

Very truly yours,

Jason Cartwright

Enclosures

Once your clients have signed a contract for a custom home, you may need to review several aspects of the building process and procedures with them in writing as shown in Figure 2.27. Issues you may want to cover in such a letter include the financing application, construction and engineering permits, and easements. Notice how the letter in Figure 2.27 explains what the client needs to do in a simple, step-by-step manner. The letter concludes by summarizing the next steps the builder will take in the process and ends the letter on a positive note.

Figure 2.27 Postcontract Letter to a Client

Jason Cartwright Builders, Inc.
101 Hilldale Court
Annandale, Virginia 20200
(703) 555-1492

June 30, 19XX

Mr. and Mrs. Winfred Scott
123 Village Parkway
Annandale, Virginia 20026

Dear Mr. and Mrs. Scott:

Thank you again for choosing Jason Cartwright Builders to build
your new home. I hope you will be pleased with our work.

Although outlined in the contract and addenda, this letter
reviews several items that we need to address within the next
week.

First, under your contract, you will need to apply for financing
by the end of next week and then notify us of which lender you
have chosen. As we discussed, you should apply for a combined
construction and permanent loan. You will also need to deliver
architectural plans and specifications to the lender after we
have reviewed and signed off on them. If you need any additional
information from us for the lender you select, please let us
know.

Before you submit a final loan application, you might want to
discuss the draw schedule with your lender. The lender may be
able to set up your construction loan draws to correspond with
our agreed-upon advances. However, of course, this is a personal
matter between you and your lender.

The permitting process also needs to begin before the start of
any engineering or construction. Jason Cartwright Builders will
purchase the construction permits. However, the property is in
your name, so you will need to sign for several of the permits.
We will arrange a meeting at the county courthouse for a convenient time to do this.

In addition, as we have discussed, the engineering firm must
first finish about 80 percent of the drainage and backfilling
before home construction can begin. The firm can then complete
the rest of the engineering work after the structure of the home
is up but before the final landscaping begins.

The engineering firm will conduct this work under a permit
separate from the construction permits. Although that firm is
responsible for this phase of the work, we should initiate the
permitting process for the engineering work, as well as the
construction work, before the former phase begins. We will be
happy to contact the engineering firm about this matter for you.

Your attorney will also need to handle the easement problem for
the utility company. This should be a routine task, but again we
should try to take care of this before construction begins.

Again, if you need any further information that we can provide
you or your attorney on this issue, please let us know.

In the next few days, we will schedule a meeting to review your
plans and specifications with you before you submit them to your
lender. At that point, we will also plan a time for you to sign
for the construction permits.

Again, please give us a call if we can help you further with any
items discussed in this letter. We look forward to starting your
new home soon and working closely with you to complete it over
the coming months.

Sincerely,

Jason Cartwright

You can turn nearly everything you do into a marketing opportunity, including
your correspondence. The prospecting letter in Figure 2.28 is similar to the letters in
Figures 2.12 and 2.18. Naturally, you need to get the Scotts' permission before you
use their name in this type of letter, but the moment you begin construction on their
new home might be a great opportunity to market your company's services to their
neighbors!

Figure 2.28 Prospecting Letter

Jason Cartwright Builders, Inc.
101 Hilldale Court
Annandale, Virginia 20200
(703) 555-1492

June 30, 19XX

Mr. and Mrs. William Mitchum
125 Village Parkway
Annandale, Virginia 20026

Dear Mr. and Mrs. Mitchum:

Your neighbors, the Scotts, have selected Jason Cartwright
Builders to build a new custom-designed home for them!

Our company specializes in building upscale homes in the newer
communities near the bay. As the president of the company, I
have 12 years of experience in the homebuilding business, all
here in Annandale.

You or someone you know may soon be interested in moving up to a
Jason Cartwright Home, and we would certainly be interested in
talking with you.

Enclosed is a brochure that describes the type of homes we build
and a list of references. Please give me a call or have your
agent call me if you are interested in our services.

Very truly yours,

Jason Cartwright

Enclosures

The letter in Figure 2.12 introduced your company to the new neighbors in a speculative building situation. You will also want to write a similar letter in a custom building situation. The letter in Figure 2.29 shows the subtle, yet important differences between the two types of letters. For example, the letter in Figure 2.29 carries no mention of an open house or discussion of early-bird discounts offered during the first few weeks of construction.

Figure 2.29 Form Letter to New Neighbors

Jason Cartwright Builders, Inc.
101 Hilldale Court
Annandale, Virginia 20200
(703) 555-1492

June 30, 19XX

Ms. Harriet Witter
128 Cedar Court
Annandale, Virginia 20029

Dear Ms. Witter:

Mr. and Mrs. Winfred Scott have selected Jason Cartwright
Builders to build their new home at 102 Cedar Court, just down
the street from your home.

Jason Cartwright Builders is a custom homebuilding company that
specializes in the kind of fine homes found in neighborhoods
like yours. We are pleased the Scotts have selected our company
to build their new home, and you are sure to be just as pleased
to have them as your new neighbors.

You may become curious as the construction progresses, but for
the sake of safety we ask that everyone--particularly children--
stay away from the construction site. Later, after the Scotts
have moved in, I am sure you will want to visit them and see
their new home.

Also, even in the best circumstances, construction can be a
noisy, messy business. However, we will try to keep the mess to
a minimum and clean up after ourselves. We will also instruct
our workers not to crank up machinery or deliver material before
8:00 a.m. If any violations of this rule occur, please let me
know immediately. If you cannot reach me, then please call our
construction superintendent, Billy Joe Spears, at 555-1861.

```
Please do not hesitate to call if you have any questions or we
can be of service to you in any way.

Sincerely,

Jason Cartwright
```

Not every letter requires the boss's signature. In fact, as part of implementing good business practices, you may want to encourage other members of your company to write letters. They are just as likely as you are to conduct business for your company that needs to be put in writing. Besides, you don't want the outside world thinking that the boss has to make every decision, do you? Whether you or another employee of your company writes the letter, a professionally run business will enforce a uniform style on all outgoing correspondence. The letter in Figure 30 comes from the construction superintendent yet follows the same format as the other letters in this chapter.

Figure 2.30 Letter Providing a Progress Report to a Client

Jason Cartwright Builders, Inc.
101 Hilldale Court
Annandale, Virginia 20200
(703) 555-1492

```
September 30, 19XX

Mr. and Mrs. Winfred Scott
123 Village Parkway
Annandale, Virginia  20026

Dear Mr. and Mrs. Scott:

Construction on your home is going well and, in fact, is right
on schedule. When building a new home, we at Jason Cartwright
Builders like to provide the homeowner with a progress report
every two weeks. This report will help you understand everything
that we are doing as we do it.

The building-area excavation went well and has been finished for
awhile. However, as you know, we experienced a short delay in
construction because the local historical commission had to
investigate a possible Native-American burial site. As builders,
we are required by law to report all such potential findings to
the commission. Fortunately, your property is not an historical
site, so construction is back on schedule.

Yesterday we set the forms and poured the concrete for the foot-
ings. These will need to cure for a few days. After that the
county will conduct a routine inspection of the footings. Then
the block masons are scheduled to begin work on October 5. I
will be on the site all day on the 5th. You may want to drop by
at that time and discuss some of the rough-grading issues that
we face after the foundation is finished.
```

```
If you have any questions, please feel free to call me at work
(555-1861) or at home (555-1912). Thank you, and it is a plea-
sure working for you.

Sincerely,

Billy Joe Spears
Construction Superintendent

cc:  Jason Cartwright
```

As the letter in Figure 2.30 indicates, normally you should include the title of the person who wrote the letter right after his or her name in the signature space. Even with a small business, this allows the person who receives the letter to understand just who has sent it.

Hopefully this chapter has now shown you a core of sample letters the beginning builder can use, as well as letters the established builder who produces speculative homes can use. The next chapter will begin to look at the correspondence needed in more advanced management situations, particularly those faced by the established builder.

Chapter 3

Letters for an Established Builder

Reports to Suppliers, Lenders, Brokers, and Others

If you think of your employees as a corporate family, then you might think of your suppliers, lenders, brokers, and others as your extended corporate family. Periodically, extended families need to communicate among themselves. For example, your lenders will want occasional copies of financial reports. Many suppliers and brokers will require regular updates on your activities. These are fair and normal requests, and you should anticipate them before they come. In fact, by anticipating these requests, you can turn them into marketing opportunities as shown in Figure 3.1.

Figure 3.1 Cover Letter When Business is Thriving

Jason Cartwright Builders, Inc.
101 Hilldale Court
Annandale, Virginia 20200
(703) 555-1492

March 15, 19X1

Mr. John Arnold
President
Arnold Lumber Company
8895 River Rapids Road
Annandale, Virginia   20031

Dear John:

Enclosed is a copy of our audited financial statement for the year ending December 31, 19XX, as requested by your credit department. Please forward the statement to them once you have had a chance to review it.

As the financial statement indicates, by keeping overhead thin and paying careful attention to detail, we actually showed a modest profit last year. This is rare for a new firm, and we hope that it is a sign of things to come.

As you and I discussed last week, together we need to explore a small increase in our line of credit to accommodate our planned growth. This increase will allow us to continue relying on Arnold Lumber as our primary supplier for framing material. I will call you later in the week to set up an appointment, so we can finalize this matter.

```
Please pass on my thanks to your staff for helping make this a
successful year. All of us at Jason Cartwright Builders look
forward to a long and prosperous relationship with your company.

Sincerely,

Jason Cartwright
President

Enclosure
```

Every business, large or small, occasionally finds itself in a slump. How your company reacts to this slump sends an important message to your suppliers, subcontractors, lenders, and other business associates. As the letter in Figure 3.2 shows, how you describe this slump in your correspondence also sends an important message to your extended corporate family.

Figure 3.2 Cover Letter When Business is Poor

Jason Cartwright Builders, Inc.
101 Hilldale Court
Annandale, Virginia 20200
(703) 555-1492

```
March 15, 19X4

Mr. John Arnold
President
Arnold Lumber Company
8895 River Rapids Road
Annandale, Virginia  20031

Dear John:

Enclosed is a copy of our audited financial statement for the
year ending December 31, 19X3, which we annually provide to your
credit department. Please forward the statement to them after
you review it.

As the statement indicates, 19X3 was financially a disappoint-
ment for Jason Cartwright Builders, although it follows several
increasingly profitable years. However, higher overhead caused
by rapidly expanding sales volume in the past three years and a
downturn in the economy left us at the year's end with inade-
quate operating income to cover fixed charges and interest.

At the same time, the company now faces several bright spots.
First, we have substantially trimmed overhead. Second, we have
taken steps to reduce construction interest expense. These two
steps alone should restore the company to profitability this
year.

Also, our sales volume is already higher than it was in 19X3,
although it is not yet back to the peak we enjoyed in 19X2. This
additional operating income should assure a successful second
half for 19X4 and allow us to end the year well in the black.
```

```
We appreciate the patience your company has shown during these
difficult months. I realize the payment delays that resulted
from our cash-flow bind have also caused your company some
difficulties. Our company should return to a timely payment
schedule by the end of the second quarter of this year.

Thank you again for your continued support for Jason Cartwright
Builders.

Sincerely,

Jason Cartwright
President

Enclosure
```

Reading between the lines of the letter in Figure 3.2, you can see that the business relationship between Jason Cartwright and John Arnold is a strong one. They have probably developed this relationship over several years with constant, honest communication such as shown in Figure 3.2. However, some updates to suppliers can take on a different tone when the business relationship is not as strong, a problem arises, or the two parties lack common goals. You could devote an entire book to the use of correspondence in those kind of business relationships. The letter in Figure 3.3 shows one way to handle those kind of situations.

Figure 3.3 Letter Updating a Supplier

Jason Cartwright Builders, Inc.
101 Hilldale Court
Annandale, Virginia 20200
(703) 555-1492

```
March 15, 19X9

Mr. Tom English
President
English Building Supply Company
798 Bay Creek Road
Annandale, Virginia  20033

Dear Mr. English:

Enclosed is a copy of the audited financial statement for the
year ending December 31, 19X8, which Jason Cartwright Builders
annually provides to your credit department. Also enclosed is
some additional information about our company.

Over the past decade with the exception of 19X3, Jason Cart-
wright Builders has enjoyed increasingly profitable operations.
During this period we have also maintained excellent and mutual-
ly profitable relations with the finest material suppliers in
the region. Our team approach to the management of our business,
which includes suppliers, brokers, subcontractors, and lenders
in our decision-making process, has met almost uniform acclaim.
```

I say "almost uniform acclaim" because the one sore spot in our supplier relationships is with the English Building Supply Company. Try as we might, we cannot seem to develop a significant line of communications with your sales staff.

To review, we first opened an account with your firm about six years ago. Since then, our business has grown considerably, and our use of your products has grown with it. During this period, your company has changed models several times without notifying us and discontinued stocking certain key hardware items. This has resulted in our having to "special order" hardware to finish several houses. On a number of occasions, we have written letters to your company about this problem and received no response. Copies of those letters are enclosed for your review.

Our clients have come to expect certain brand-name hardware items that you carry, so we are very much interested in developing a closer working relationship with your company. Can you and I meet to discuss how our companies can work more closely together in the future? At your convenience, please give me a call and we will arrange a meeting to address this matter.

Sincerely,

Jason Cartwright
President

Enclosures

Besides providing suppliers, lenders, brokers, and others with updated financial statements as shown in Figures 3.1 through 3.3, you may want to keep them up-to-date on other business activities. Newsletters are useful, cost-effective ways of doing that. Readers of periodic newsletters begin to feel part of the corporate family, and that is exactly how you want your suppliers, lenders, brokers and other business associates to feel. Until recently, producing a high-quality newsletter was time consuming. However, today many word-processing and desktop-publishing programs can quickly produce simple but professional-looking newsletters. Figure 3.4 shows a one-page newsletter that you can easily produce with word-processing or desktop-publishing software.

Figure 3.4 Newsletter to Suppliers, Lenders, Brokers, and Others

Jason Cartwright Homes

Jason Cartwright Builders, Inc.
101 Hilldale Court
Annandale, Virginia 20200
(703) 555-1492   *June 19XX*

Jason Cartwright Builders Celebrates Its Third Year with a Gala Event!

Jason Cartwright Builders will celebrate its third year in business with a gala party on Saturday, July 13. In honor of our new Tudor line of homes, Dr. and Mrs. Russell Massey have graciously allowed their new home, the first of this line, to be the site of this all-day event. The Massey home is located at 14 Ridge Crest Drive, right on the 15th fairway of the new Golden Springs Golf Club.

The party will begin at 12:00 noon and continue until dusk. We will offer tours of the new home and provide entertainment and refreshments throughout the afternoon. Agents are invited to bring potential clients, and representatives of the Golden Springs Golf Club will give golf-cart tours of their fabulous facilities.

A special word of thanks goes to Herb Johnson Catering. In addition to this gala, that company has handled several functions for us in the past and always does a superb job. So, keep Herb Johnson in mind for your next catered event.

July 4th Holiday Notice

Jason Cartwright Builders will close for the Independence Day celebrations on Thursday and Friday, July 4 and 5. Subcontractors and suppliers who are usually paid on Friday will be paid on Wednesday, July 3, through the normal means of distributing checks. Any check not picked up that day we will mail that evening.

Jason Cartwright Builders wishes all of its suppliers, lenders, brokers, subcontractors, and their families safe and happy Independence Day celebrations.

New Homeowners!

Mr. and Mrs. Lauren Johnson have become the newest buyers of a Jason Cartwright home. The Johnsons' new home will be a Georgetown-style, two-story house with a custom-designed kitchen and swimming pool. Carl Green is the architect and Forsythe-Jenkins and Associates will handle the interior decorating. Construction on this new home begins in Arbor Forest West on July 1, with a tentative completion date of November 27.

A special thanks goes to Audrey Tewes of Newharbor Realty, who is handling the Johnson contract. This is Audrey's second sale of a Jason Cartwright home, and we are sure to see more of her in the future.

You may have other occasions for contacting your broker besides a periodic company newsletter as shown in Figure 3.4. Most brokers keep close tabs on the work of agents in their firms. When agents find a source of repeated sales or sales leads, the broker will encourage them to share that information with others. As a builder, letting a broker know that his or her agents are repeatedly selling your homes is in your best interest. The broker has an incentive to encourage other agents in the firm to work this source of repeat business. Use every opportunity to put your name in front of the broker in a positive light. As Figure 3.5 shows, an annual report is a clear opportunity to do this.

Figure 3.5 Letter to a Broker

Jason Cartwright Builders, Inc.
101 Hilldale Court
Annandale, Virginia 20200
(703) 555-1492

January 15, 19X5

Ms. Anne Baxter
President
Baxter Realty
101 Bayside Boulevard
Annandale, Virginia   20035

Dear Anne:

Thank you for the enthusiasm your agency has shown toward Jason Cartwright Builders in 19X4. As we close the books for 19X4, you and I will want to reflect on the sales our two companies have shared during this past year. Anne, three of your agents were responsible for 50 percent of Jason Cartwright home sales during 19X4.

Marilyn Duke handled three sales: the Huntley home in January, the Myers home in July, and the Davidson home in August. David Barter wrote the Harrison contract in February, the largest home we have sold to date, which closed last month. Finally, Susan Saris handled the Williams job, which we both know was tough to finance. All three of these agents deserve a pat on the back for the professionalism, expertise, and determination they showed in closing these deals.

In recent months, you and your agents have worked hard to increase the number of large, custom home sales handled by Baxter Realty. If you are interested, Marilyn and I would like to make a presentation on selling custom homes at one of your upcoming sales meetings. Presenting the custom builder's perspective might help to stimulate additional activity in this area. Please give me a call if you think this is a good idea.

```
Again, thank you for the enthusiasm your agency has shown toward
Jason Cartwright Builders. We look forward to working with
several of your agents in the near future.

Sincerely,

Jason Cartwright
President
```

As with most thank you letters, Jason tries to kill two birds with one stone in Figure 3.5. Thanking the broker for previous business gives him the opportunity to open the door for future business.

Personnel Management

As soon as you hire one employee, you become involved in personnel management. If you do not believe this, call your local department of labor or employment security commission. They will gladly explain it to you in great detail. Regardless of the size of your operation, effective written communications with your employees is key to their morale and productivity. It is also important in forging good employer-employee relations.

Many larger building companies have formal procedures developed over many years for handling certain types of written communications with employees. On the other hand, some mom-and-pop operations may take employee communications for granted, assuming that the company's family atmosphere will foster more direct, informal communications. Since statistics show that the largest increase in total employment in the United States over the last two decades has occurred in mom-and-pop businesses, clearly the need for formal, written communications with employees in small businesses is real.

Regardless of the size of your company, you should provide every prospective employee with a clear set of the company's goals and objectives. After all, you do not want to waste time and money hiring an employee only to find that person does not want to work for a company like yours. A larger firm may have a formal brochure that describes its goals and objectives. A smaller company with little turnover might have a form letter preprinted to accompany each job application as shown in Figure 3.6.

Figure 3.6 Letter Explaining a Company's Goals and Objectives

Jason Cartwright Builders, Inc.
101 Hilldale Court
Annandale, Virginia 20200
(703) 555-1492

```
Dear Prospective Employee:

On behalf of Jason Cartwright Builders, thank you for your
interest in employment with our company.

Jason Cartwright Builders was formed in 19XX to build large
custom homes in what were then the newer neighborhoods around
the bay. That year we proudly built four homes, and we take that
same pride in every home we build today. Last year, the company
```

built over 20 large custom homes in northern Virginia and
Maryland.

In addition to building custom homes, in recent years we have
opened a small retail store, Cartwright Hardware Company, and
completed our first development project, Newcity Development.
The company has also become involved in two commercial ventures.
However, we still keep "Builders" in the company's name to
emphasize our core business.

Generally, Jason Cartwright Builders has a policy of considering
current employees for any openings. As the company grows, we
anticipate openings in field supervision and management and will
make every effort to fill these positions with current employees. However, we will also seriously consider all qualified
applicants for any open positions.

Part of the company's success through the years comes from a
low-overhead philosophy. If you visited our offices, you
probably noticed that the office staff and facilities are
functional, with the most up-to-date equipment and technology
available. However, as you know, homebuilding is a cyclical
business. In response to that, we keep overhead lean so we can
move quickly to tighten our belts or to take advantage of
available opportunities. All of our staff is adaptable and
willing to respond to market demands.

At Jason Cartwright Builders, the customer may not always be
right, but the customer is always the customer. This company is
a totally customer-oriented business. If you think of yourself
as a customer-oriented person, then Jason Cartwright may be the
company for you.

Again, thank you for your interest in employment at Jason
Cartwright Builders. We wish you the best of luck in all your
future endeavors.

Sincerely,

Jason Cartwright
President

If you think the letter in Figure 3.6 sounds like a selling job, you are right. You want to sell your company's goals and ideals to new employees, even before they join the company. Successful companies work hard at making sure every employee sings from the same songsheet.

As shown in Figure 3.7, you should make a letter offering a job to someone a bit more formal and personalized than the letter shown in Figure 3.6. In many instances, this letter can be construed as a binding legal document. Thus, you want to avoid making promises or even hinting at promises that you cannot keep down the road. For example, the following phrases might be dangerous if contained in a letter to a prospective employee:

- "Our company has a firm commitment to permanent employment with no layoffs."
- "After six months, you can reasonably expect a promotion and a pay raise."
- "We *always* promote from within."

- "While two weeks of vacation is our policy, we rarely have a problem with an employee taking a day or two off now and then, as long as the work is caught up."

When in doubt, try to imagine yourself explaining every word of the offer letter while sitting on a witness stand.

Figure 3.7 Formal Letter Offering a Job

Jason Cartwright Builders, Inc.
101 Hilldale Court
Annandale, Virginia 20200
(703) 555-1492

September 9, 19XX

Mr. John Dunlap
108 Cedar Terrace
Annandale, Virginia   20037

Dear Mr. Dunlap:

With this letter, Jason Cartwright Builders offers you a job as a truck driver, working for our construction superintendent, Billy Joe Spears. Your first day on the job will be September 15, 19XX.

As explained to you in your interview, this job is an hourly position, normally working 40 hours per week with occasional overtime. Your starting base pay will be $5.75. You will be on probation for the first 90 days.

Enclosed is a copy of our brief employee handbook. This describes all of our company rules and regulations as well as the fringe benefits you can expect. This book also explains what will be expected of you during your probationary period.

Please read the employee handbook carefully. If you have any questions about what it covers, my assistant, Susan Kirkland, handles all of our personnel matters and will gladly explain it to you.

If you accept this position and the terms outlined in this letter, please sign the two enclosed copies of the letter, return one signed copy to us, and retain one copy for your records.

Once you have returned a signed copy of the letter to us, please call Susan by September 13. She will then give you instructions for your first day on the job on September 15.

```
Again, we appreciate your interest in Jason Cartwright Builders
and look forward to your joining the company.

Sincerely,

Jason Cartwright
President

Enclosures

I agree to the terms outlined in this letter and have received a
copy of the employee handbook.


_____
John Dunlap              (Date)
```

The letter in Figure 3.7 refers to an employee handbook. Every company should have one. Even small companies with only a few employees can write up a few pages. The employee handbook should outline such things as your vacation and sick-leave policies and group insurance, retirement, and other fringe benefits. Additionally, the handbook should outline company policies about such issues as on-the-job safety and injuries, disability leave which legally also applies to maternity leave, absenteeism, tardiness, and on-the-job alcohol and drug abuse.

Use great care and thought when writing your employee handbook regardless of how simple or elaborate it becomes. Many potential sources of advice are available to you in this process, including your insurance agency, attorney, or local chamber of commerce. If a nearby college has a business department, then the assistance of a faculty member in personnel management would be helpful. However you do it, you need to spell out all of your company's rules and policies in writing and give them immediately to every new employee. This important step will save you time, aggravation, and money down the road.

As mentioned in Figure 3.7, most companies, whether large or small, have a probationary period for new employees. This period is usually 30 to 90 days and is followed by a formal evaluation. In reality, the probationary period gives the employer an opportunity to conduct a 90-day evaluation of the new employee, which in turn gives the employee a chance to see how well he or she is meeting the company's expectations.

Many large companies use formal evaluation forms developed by teams of personnel experts. However, small and medium-sized companies can usually get by with a memo, summarizing the results of an evaluation meeting. For an example of such a memo, see Figure 3.8. As shown in this figure, some companies like to have the employee sign or initial the evaluation as an acknowledgment that he or she has read and understood the details covered.

Figure 3.8 Memo Containing an Employee's Evaluation

Jason Cartwright Builders, Inc.
101 Hilldale Court
Annandale, Virginia 20200
(703) 555-1492

December 12, 19XX

Memorandum to: John Dunlap _____(initial)

From: Billy Joe Spears, Construction Superintendent

Subject: Probationary Period Evaluation

As a normal part of employment, every individual at Jason
Cartwright Builders receives a probationary period evaluation.
This is a chance for you and your supervisor to discuss your
performance to date, including your strengths and those areas
requiring improvement.

Please read this entire memo and then initial it in the space
provided by your name to show that you have read and understood
it. Your supervisor will then discuss each point with you and
answer any questions you may have. Your next evaluation will be
on or about the one-year anniversary of your employment, and you
will have evaluations about every year thereafter.

1. Over the past three months, you have appeared to be an eager
learner and have worked hard to improve your job skills. The
company appreciates this and wants to encourage you to continue
this behavior in the future.

2. You have arrived late to work five times in the last three
months (9/15, 9/22, 10/15, 11/17, and 11/18). Other people in
the field count on you, and you create extra work for them when
you are not there. We need to see an immediate and consistent
improvement in this area or further disciplinary action will be
taken, which may include termination of your employment.

3. Both insurance and government agents periodically conduct
safety inspections of our job sites. To prepare for these inspections, we conduct occasional spot checks on all employees.
During these spot checks, you were observed on two occasions not
wearing goggles (9/13 and 10/1) and once not wearing a face mask
(10/1) when they were required. These rules were made for your
own safety, and we must enforce them. Any further failure on
your part to follow safety rules may result in further disciplinary action and may be grounds for dismissal.

Again, it is important that you take this evaluation in a positive spirit. Generally the company is pleased with your performance, and we all want to see you grow with Jason Cartwright
Builders. However, it is also important that you follow the
company's procedures. If you have any questions about this evaluation or your performance in general, please do not hesitate to
ask me at any time.

Notice the evaluation memo in Figure 3.8 begins by mentioning the one positive thing the employee was doing and ends on a positive note in looking toward the future. These are important motivational tools that should not be ignored. The memo also refers to specific instances when the employee violated the company's rules. Every employer has warned an employee about something at one time or another. However, if you do not put the warning in writing, you have no proof that you ever issued it. Later, if you have to discipline or dismiss the employee, you will have no evidence to stand on in court. Figure 3.9 shows an example of an employee warning letter.

Figure 3.9 Employee Warning Letter

Jason Cartwright Builders, Inc.
101 Hilldale Court
Annandale, Virginia 20200
(703) 555-1492

November 9, 19XX

Memorandum to: Cindy Weatherall

From: William McConicle, Accounting Manager

Subject: Tardiness

According to the company's records, you have arrived late to work 14 times in the past 60 days, ranging from 10 minutes to three-and-a-half hours late. On average, you have come late to work every third day and have missed a total of over seven hours of work because of tardiness, which is nearly a complete work day.

Naturally, you will not get paid for the time missed. In addition, your absence puts an undue burden on the other employees. Your work still has to be done, and when you are not here, it falls on the shoulders of the other employees. Obviously, this is unacceptable.

In looking back over the company's records, I note that your tardiness seems to have started around the first of September. However, if immediate and consistent improvement is not seen with respect to your tardiness, further disciplinary action will be taken, which may include termination.

Cindy, you are a valuable employee, and we certainly do not want to lose you. However, permitting you to continue arriving late for work would be unfair to the other employees. Please work with us to resolve this issue.

In addition to issuing a warning letter as shown in Figure 3.9, every manager has to dismiss an employee every now and then. Often this is done verbally ("Get off this property now or I'm calling the security guards!"). However, you need to follow up a verbal dismissal with a letter like the one shown in Figure 3.10—if nothing else, so you have a record for your files.

Figure 3.10 Letter of Termination

Jason Cartwright Builders, Inc.
101 Hilldale Court
Annandale, Virginia 20200
(703) 555-1492

October 4, 19XX

Mr. Jeremy Roberts
101-C Mockton Court
Annandale, Virginia   20040

Dear Mr. Roberts:

This letter confirms that your employment with this company is terminated effective today. We will mail your final paycheck to your home address of record on the next regular payday.

For the record, you were driving truck number 2 early this morning, loaded with building material destined for 677 Northridge Lane where this company is building a home. John Johnson was in the cab of the truck with you.

Around 8:30 a.m. Ms. Susan Hancock, one of the residents of Northridge Lane, saw truck number 2 proceeding in an erratic manner down the street. You stopped the truck in front of 677 Northridge, and Ms. Hancock saw you and Mr. Johnson "fall out of the truck, the passenger landing in the shrubbery and the driver falling face first onto the pavement."

Thinking you were ill, Ms. Hancock ran to your aid. She reported that you and Mr. Johnson both "stank of alcohol, and two empty wine bottles rolled out of the cab onto the pavement." The construction superintendent came over about that time, helped you and Mr. Johnson sit up on the sidewalk, and called the county sheriff. Later today around 11:00 a.m. you were both arrested on charges of public drunkenness.

Jason Cartwright Builders has the following absolutely firm company policy, which is stated in the employee handbook that you received your first day on the job: "Any employee suspected of being under the influence of alcohol or narcotics while at work will be asked to leave the company premises, and if it is later proven that the employee was under the influence of either, the employee will be terminated. There are no exceptions."

In this situation, I have no other choice but to terminate your employment.

Sincerely,

Jason Cartwright
President

Finally, most letters you write to or for an employee will not contain bad news like the letter in Figure 3.10. Frequently you will have the pleasure of writing a letter that recommends a current or former employee for another job. You walk a fine line with these types of letters. You want to sound complimentary but not mushy. Honesty is always the best policy. But how do you handle disclosing some obscure but perhaps relevant fact that could affect the new employer if he or she hires this person? "Alice is a nice person, but don't let her near sharp knives" or "Bill is the sharpest accountant I've ever met, but he has this 'thing' for cash...."

The level of honesty you should strive for in a letter of recommendation is a tough call. For that reason, you may want to keep your letter more general than specific. You may also want to avoid a great deal of detail in the letter. In fact, the shorter the letter of recommendation is, the better! When in doubt, try to recall some letters of recommendation that you have received over the years and use them for guidance. Figure 3.11 shows one example of such a letter.

Figure 3.11 Letter of Recommendation

Jason Cartwright Builders, Inc.
101 Hilldale Court
Annandale, Virginia 20200
(703) 555-1492

April 10, 19XX

Ms. Toni Fenwick
President
Fenwick Realty
Post Office Box 8
Annandale, Virginia   20041

Dear Ms. Fenwick:

I strongly recommend Sally Satterfield for the position of accounts payable manager with your organization. As you know, Sally has worked with Jason Cartwright Builders for five years as a senior accounts payable assistant. During that period she has performed her job with excellence. If we had an opening into which we could promote her, we would not hesitate to consider her for it.

Several aspects of Sally's performance may not have come out in her resume and interview. Sally is an extremely dedicated and hard-working employee. She is one of the first people here every morning and is usually the last to leave in the evening.

On her own time Sally has expanded her education in many ways. She took the initiative to become trained as a systems manager on our computer system, so we would have a backup if our regular systems manager was ever gone for an extended time.

She has also become extensively familiar with the accounting software we use and has recommended several improvements to the system that the authors have accepted. In short, Sally is a valuable employee, and you would be lucky to get her.

At Jason Cartwright Builders we are all sorry to see Sally leave
and will miss her. At the same time, everyone here is glad that
she can expand her horizons through this job change. We wish her
the best of luck in her work at Fenwick Realty and all her
future endeavors.

Sincerely,

Jason Cartwright
President

In summary, personnel matters are becoming more and more complicated for even the smallest company. Certainly, an entire book can be written on personnel correspondence. In the preceding section, Figures 3.6 through 3.11 have hopefully given you a glimpse at some ways to handle this type of correspondence.

Dealing with Insurance Companies

Besides providing regular written reports to suppliers, lenders, brokers, and others, and putting personnel matters in writing, established builders have to deal with a variety of other companies on a daily basis, including insurance companies. Your dealings with insurance companies can easily become complicated, time consuming, and costly—all the more reason to put these dealings in writing. The letters in Figures 3.12 and 3.13 show two of the more common situations you will encounter in dealing with insurance companies. Again, they give you only a glimpse of how to handle these and other similar types of correspondence.

Figure 3.12 Letter Contesting a Premium Increase

Jason Cartwright Builders, Inc.
101 Hilldale Court
Annandale, Virginia 20200
(703) 555-1492

March 30, 19XX

Mr. Robert Foster
Foster and Gaines Insurance
12 Hardtop Road
Annandale, Virginia 20042

Dear Bob:

Jason Cartwright Builders has dealt with your firm since the
first day we opened our doors. Since then you have been our only
insurance agents and your company has provided us with excellent
service. In turn, as our company has grown, we have brought your
company increasingly larger amounts of steady business.

Regretfully, the premium increase in our most recent workers'
compensation insurance contract is unacceptable. We are fully
aware that the entire building industry has experienced a shift
in liability problems over the last few years. However, our
company is not the whole industry.

In fact, Jason Cartwright Builders has taken major steps, many
with the help and encouragement of the insurance carrier, to
minimize on-the-job accidents, time lost from accidents, and
size of accident claims. As part of those major steps, we have--

*   assigned loss control and safety as a half-time duty to a top
manager

*   instituted monthly mandatory safety briefings for all
employees

*   rewritten the safety section of our employee handbook

*   started a monthly safety newsletter, which is posted at every
job site

*   rewritten many other company policies regarding safety and
work habits

As a result of those steps, we have cut our quarterly number of
on-the-job accidents in half, cut total lost time from accidents
by two-thirds, and increased days between accidents four times.
Although the average claim size is skewed up a bit from that one
tragic accident earlier this year, the median claim size has
decreased.

In response to this positive activity on our part, the insurance
carrier has increased our premium by 31 percent. Under the circumstances, you have to agree that this increase is unreason-
able.

Bob, we would appreciate your looking into this matter and calling us with some suggestions on how together we can resolve it
to everyone's satisfaction. We look forward to hearing from you
in the next week.

Sincerely,

Jason Cartwright
President

Notice how the letter in Figure 3.12 separates out important points with bullets
for emphasis. If the letter in Figure 3.12 does not produce positive results, you may
need to write a letter shopping for another insurance company as shown in Figure 3.13.

Figure 3.13 Letter Shopping for Another Insurance Company

Jason Cartwright Builders, Inc.
101 Hilldale Court
Annandale, Virginia 20200
(703) 555-1492

May 12, 19XX

Ms. Sherree Cato
The Elliott Agency
4565 Elliott Building
Annandale, Virginia   20043

Dear Ms. Cato:

You were referred to me by my good friend, Jeffrey Smythe of
Smythe Builders. He praises the work you and your agency have
done in handling his insurance over the past few years and
suggests that we talk about coverage for Jason Cartwright
Builders.

This company was formed in 19XX to build large custom homes in
what were then the newer neighborhoods around the bay. That year
we proudly built four homes and take that same sense of pride in
every home we build today. Since then, the company has built
homes in several other communities. In fact, last year we built
over 20 homes in northern Virginia and Maryland.

In addition to building custom homes, we have opened Cartwright
Hardware Company and have completed our first development project called Newcity Development. The company has also become
involved in two commercial ventures. However, we keep "Builders"
in the company's name to emphasize our core business.

Over the past few years, Jason Cartwright Builders has taken
major steps to minimize on-the-job accidents, time lost from
accidents, and size of accident claims.

As part of those steps, we have assigned loss control and safety
as a half-time duty to a top manager; instituted monthly mandatory safety briefings for all employees; rewritten the safety
section of our employee handbook; started a monthly safety newsletter, which is posted at every job site; and rewritten many
other company policies regarding safety and work habits.

As a result of taking those steps, the company has cut the quarterly number of on-the-job accidents in half, cut the total lost
time from accidents by two-thirds, and increased the days between accidents four times.

```
By July 1 Jason Cartwright Builders would like to change insur-
ance carriers. Before making a decision, we plan to meet with
representatives of two or three top agencies. Please call to
schedule an appointment with us to discuss the specifics of
insurance coverage for our company. We look forward to hearing
from you soon.

Sincerely,

Jason Cartwright
President
```

Did the letter in Figure 3.13 sound familiar? The second and third paragraphs contain material from the letter in Figure 3.6 on the company's goals and objectives given to prospective employees. Paragraphs four, five, and six contain material from the previous letter in Figure 3.12. That is one of the great advantages of word processors. You have the option of saving and reusing well-thought-out writing again and again.

Engaging Professional Services

In addition to selecting and dealing with insurance companies, an established builder will often hire another firm to perform special services on an ongoing basis. This will include the accounting firm that does your taxes, the janitorial firm that cleans your model homes or offices, and even an auto repair shop that maintains your company vehicles.

A letter of engagement is a formal go-ahead for such a firm to start work. For an example of such a letter, see Figure 3.14. This kind of letter is a much weaker document than a contract. In fact, based on a letter of engagement, a firm may bring you a contract to back up the relationship. Nevertheless, the letter of engagement allows the firm that you hire to begin representing your interests in such matters as filing tax returns, receiving summons, and changing the oil in all of the pickups.

Figure 3.14 Letter Hiring an Attorney on Retainer

Jason Cartwright Builders, Inc.
101 Hilldale Court
Annandale, Virginia 20200
(703) 555-1492

```
July 17, 19XX

Mr. Timothy Stirren
Attorney-at-Law
400 Executive Plaza, Suite E
Annandale, Virginia  20044

Dear Mr. Stirren:

Please accept this letter as a formal engagement of your
services as attorney for Jason Cartwright Builders, Inc.,
effective July 31, 19XX. The Board of Directors voted on and
approved this matter in its monthly meeting on July 16, 19XX.
```

```
The terms of your letter dated July 6, 19XX, are accepted in
their entirety, and that letter is attached to and made a part
of this letter of engagement. Further, a retainer against actual
billable services of $300 per month will be paid at the first of
each month, beginning August 1, 19XX.

On behalf of the company, we are delighted to be working with
you and look forward to a long and successful relationship with
your firm.

Sincerely,

Jason Cartwright
Chairman of the Board
and Corporate Secretary

Attachment
```

In Figure 3.14 Jason Cartwright's title in the letter is very important. This letter actually reports the actions of the company's board of directors, and Jason is reporting those actions in his capacity as secretary. Even if he is the only member of the board, he is still acting as the secretary of the board.

Chapter 3 has assumed that the firm sending the letters is established and stable, rather than young and growing. However, just because a business is established does not mean that it cannot grow. Many older companies grow either in periodic spurts or slowly and continuously over time. The next chapter will look at the correspondence needed by an ongoing business that still wants to grow and expand.

Letters for a Growing Business

Business growth is like a garden. You have to plan for, work at, and carefully cultivate it. If you ignore growth or assume it will happen on its own, it will turn into a patch of weeds. If you tend growth properly, it may turn out the way you want it to. Since part of that tending includes correspondence, Chapter 4 focuses on some of the letters needed by a thriving business that wants to grow and expand. Most of the letters in this chapter are also useful for businesses of any size, including smaller and newer builders.

Extending Lines of Credit or Seeking Alternative Financing

The search is on for new lines of credit! How many builders in your neighborhood are satisfied with their current credit arrangements? Are you? Figures 4.1 through 4.4 show examples of letters that an experienced, thriving builder might use to expand or change credit arrangements.

Figure 4.1 Letter to a Construction Lender

Jason Cartwright Builders, Inc.
101 Hilldale Court
Annandale, Virginia 20200
(703) 555-1492

```
January 5, 19X1

Ms. Delila Barr
Vice President
State Street Bank and Trust
3 Main Street Tower
Annandale, Virginia  20045

Dear Ms. Barr:

Chatting with you last night at the home builders meeting was a
pleasure. Everyone is looking forward to your firm's entry into
construction lending, and no doubt many eager builders will call
you over the next few days.

Jason Cartwright Builders is no exception. In this market, we
are always looking for new sources of construction loans and are
interested in hearing more about what your bank has to offer.
From our side, we can offer a prosperous, growing company with a
nearly unbroken string of profitable years.
```

```
Over time we have developed a conservative management style and
an aggressive procustomer orientation that have served us well.
Enclosed is a brochure, recently developed by our marketing
consultants, that will describe more of our business philosophy
and sales orientation.

Our company is also mildly diversified, but only to an extent
that enhances the core business of building custom homes in new,
upscale neighborhoods throughout northern Virginia and Maryland.

If you feel as we do that State Street and Jason Cartwright
Builders would make a good business fit, we would like to submit
a more comprehensive presentation of our business along with a
thorough loan application. With your permission, I would like to
call you within the next week to start that process.

Thank you, and we look forward to seeing you soon.

Sincerely,

Jason Cartwright
President

Enclosure
```

This is a rather open-ended letter. Jason does not ask for specific loan commitments, nor is he anxious to rapidly establish a line of credit so he can begin work on his first house. He is simply developing a new source of credit by beginning a process that may take months to finalize. If his business is to continue to grow and prosper, he must continually develop these new sources of credit.

Like relationships with lenders, you need to carefully develop and cultivate relationships with suppliers. Occasionally you may even need to end a relationship with a supplier. Figure 4.2 shows a letter to a supplier in one such relationship.

Figure 4.2 Letter to a Supplier

Jason Cartwright Builders, Inc.
101 Hilldale Court
Annandale, Virginia 20200
(703) 555-1492

```
February 8, 19XX

Mr. Kenneth Farwold
President
Farwold Building Supply Company
809 Avenue of the Pines
Baltimore, Maryland  30046

Dear Mr. Farwold:

Dealing with your company has been extremely convenient and your
prices very reasonable over the years. In fact, so much so that
Jason Cartwright Builders currently purchases large quantities
of trim material from four of your stores in Maryland.

At three of those stores, the commitment to quality and service
matches our own. However, the quality of material at the Upper
```

```
Marlboro store has drastically declined over the past nine
months.

On several occasions during this period, we have spoken with the
Upper Marlboro store manager and his regional supervisor about
this issue. In a further attempt to address this issue, we have
also corresponded with them frequently. Copies of the letters
are enclosed.

Although we have received form letters in reply saying that the
Upper Marlboro store is taking steps to stop the decline in the
quality of material, it still continues. Additionally, the store
manager has been extremely slow to replace damaged or inferior
material. This in turn has caused serious and costly delays in
the affected homes.

Effective today we regret that we must suspend purchases from
the Upper Marlboro store and cancel all outstanding purchase
orders. We will accept any material actually being shipped as of
today, subject to our previously stated quality requirements.

In addition, in the next few weeks, we will review our entire
relationship with Farwold and consider finding a supplier who
can more consistently meet our quality and service requirements.

Mr. Farwold, the current situation does not have to be irrevo-
cable. Let's put our heads together and see if we can find a way
to work this matter out. At your earliest convenience, please
give me a call.

Sincerely,

Jason Cartwright
President

Enclosures (5)
```

In Figure 4.2 the "(5)" after "Enclosures" refers to the number of items enclosed with the letter. This is a convenient check for the person receiving the letter to ensure he or she got the entire package you sent. The letter in Figure 4.2 also explains the popularity of fax machines. If you were Kenneth Farwold, would you want to wait for the regular mail to find out that you were about to lose all of this business?

As businesses grow and expand, some new relationships develop, even among old partners. The letter to a subdivision developer in Figure 4.3 shows just how this can happen.

Figure 4.3 Letter to a Subdivision Developer

Jason Cartwright Builders, Inc.
101 Hilldale Court
Annandale, Virginia 20200
(703) 555-1492

November 2, 19XX

Ms. Amelia Kent
Kent Properties
1100 Dixon Line Drive
Annandale, Virginia 20047

Dear Amelia:

Over the years Jason Cartwright Builders has been one of your
best customers. Together our companies now have a great
marketing opportunity in your new community, Annandale Hills.

As you recall from our meeting last week, we at Jason Cartwright
Builders are very enthusiastic about this new development. In
fact, we plan to build over half of the homes in this
neighborhood.

Given those plans, we believe that both of our companies would
benefit from having a model home located near the subdivision
entrance. Naturally, this will require some commitments from
both of us.

From our side we propose to build, furnish, and staff the model
home with sales agents. Jason Cartwright Builders will also pay
for maintenance costs, utilities, signage, parking, and other
details. In addition, our company proposes to purchase two lots,
side by side, and use one of them for parking.

When the sales center has outlived its usefulness, we will
renovate and sell it. At that time we will also convert the
parking lot into a buildable lot for resale to one of our custom
home clients.

All we ask from Kent Properties is two things. First, we want
your company to carry a 100 percent lot mortgage on both of
these lots until they are sold to homeowners. Our attorneys can
work out the details of this arrangement.

Second, we want a commitment from you that you will not grant
any other builders similar model home rights in phase one of the
property. In the remaining phases, we will waive this require-
ment.

Today's market requires an aggressive posture. This sales center
will substantially increase our sales volume in Annandale Hills
and thus is worth a significant commitment of resources. With
the long-standing relationship between our two companies, we
felt you should receive first crack at this proposal.

```
If you want to join in this commitment, please call me as soon
as possible. Thank you for your thoughtful consideration of this
matter.

Sincerely,

Jason Cartwright
President
```

The letters in Figures 4.2 and 4.3 reveal the importance of correspondence in effective negotiating. In Figure 4.2, the letter sets the stage for negotiation by emphasizing that the company values its ongoing relationship with the Farwold Building Supply Company. Next it clearly states the problem and what Jason has done to address the problem—with little success. The letter then states the more drastic steps Jason plans to take but leaves an out for Mr. Farwold to resolve the problem amicably. Chances are that Mr. Farwold will try to do that, in part because of the way Jason has written this letter.

In the same way, the letter in Figure 4.3 introduces an idea for a joint project by reminding Amelia that Jason Cartwright Builders has been one of her best customers over the years. Next the letter describes Jason's idea, its potential benefits for both companies, and what commitments Jason Cartwright Builders will make. The letter then asks for two specific commitments from Kent Properties. Again, in part because of the way Jason has presented the idea in his letter, Amelia will probably agree to at least part of the deal.

Putting things in writing really is the best approach when seeking lines of credit or alternative sources of financing, resolving business differences, or embarking on a business arrangement with another company. As Figure 4.4 clearly shows with respect to joint ventures, how you put things in writing is also critical in these situations.

Figure 4.4 Letter Proposing a Joint Venture

Jason Cartwright Builders, Inc.
101 Hilldale Court
Annandale, Virginia 20200
(703) 555-1492

```
November 15, 19XX

Mr. Wallace Marlowe
Marlowe Homes
158 Shirley Street
Annandale, Virginia 20048

Dear Wally:

On November 12 Amelia Kent made a commitment to allow Jason
Cartwright Builders to locate a sales center in her new
subdivision, Annandale Hills. However, she prefers to have at
least two substantial builders share in building out phase one.

Since your plans for the Cliffs of Annandale are now on hold,
would you be interested in a joint sales center in Amelia Kent's
new community?
```

```
Enclosed is a copy of my original letter to Amelia. In it, I
asked that no other builders have model homes in phase one. With
the combined strength of our two companies, we could ask for--
and reasonably expect--total exclusivity for at least the first
six months.

If you are interested, we could consider building one model home
as a joint venture, with a style that combines our two compa-
nies' approaches to design. Your agents would use one-half of
the sales center and ours, the other half. The agents for both
companies would share common areas such as the conference room,
kitchen, and receptionist area.

In addition, we could set up a joint fund, co-own the center and
the parking lot, and share the proceeds when we convert the
center into a home.

Wally, let's sit down in the next few days and discuss this
idea. If we go in on this idea together, we can both benefit
and successfully sell Amelia Kent on it. At your earliest
convenience, please give me a call to schedule a meeting.

Sincerely,

Jason Cartwright
President

Enclosure
```

Can you see how putting the idea for a joint venture in writing actually helps to sell the idea to a potential partner? Keep that in mind every time you pick up the telephone to propose an idea to a business associate.

Expanding and Reorganizing the Staff

Growth involves more than seeking extended lines of credit or alternative forms of financing. It may also involve expanding and reorganizing your staff. At some point when a business grows, you may have to hire or promote an employee. Other employees may stay in their same jobs, and some may even find themselves below where they were before.

With all the change that growth brings, a company can begin to experience growing pains. Even in the smallest of companies, you can use memos to manage that change effectively. By clearly announcing the changes and the reasons for implementing them in writing, you avoid any rumors and uncertainty that are likely to occur. Stressing certain things like a family atmosphere and teamwork will also help to soothe any hurt feelings that may arise.

Figure 4.5 Memo on Reorganization

Jason Cartwright Builders, Inc.
101 Hilldale Court
Annandale, Virginia 20200
(703) 555-1492

November 15, 19XX

Memorandum to: All Staff

From: Jason Cartwright

Subject: Company Reorganization

As a company grows and changes, so do the types of jobs that
need to be done. For example, when I started this company, I was
the personnel manager. Later, Susan Kirkland took on the role of
half-time personnel assistant and half-time accounts payable
manager. Later still, Sam Smithers became full-time personnel
and safety manager. Who could have predicted those changes a few
years ago when this whole operation was run from my pickup
truck?

Today there is only one of me and a lot of you. When President
Eisenhower left office, he complained he spent so much time
working on the urgent things that he never had time for the
important things. The company has very talented employees, and
with this reorganization, I am giving a lot of you the urgent
tasks, so I can spend more time on the important ones.

Attached is a chart that shows our new organization as recommended by our consultants. You will see your name in one of the
boxes. As usual, remember that each of you individually has an
important contribution to make to the success of this company.

With the new organization, you can now go to others in the
company besides me to resolve most issues. So, for example, if
you have a personnel issue, please take it directly to Sam
Smithers. With an accounting question, you can go directly to
Mary Freeman for an answer. If you have an issue regarding
suppliers and subcontractors, see Jack Perry. Once you have done
that, if you still have not resolved your problem, my door is
always open.

Like anything else we do, this reorganization will only work if
together we all support it as a team. Over the next few weeks I
will meet with each of you to discuss the reorganization. If you
see some area that needs changing or some part of this new
structure that you think will not work, please let me know. I
cannot promise to change everything, but I do promise to listen.

As always, thank you for your dedication and hard work. I want
you to know that I appreciate it.

Like the reorganization described in Figure 4.5, you also need to handle promotions carefully, particularly if more than one person in the company has applied for the promotion. Announcing the promotion in writing is always the best approach. Be sure to stress the unique talents or abilities of the person who was promoted.

Figure 4.6 Memo Announcing Promotions

Jason Cartwright Builders, Inc.
101 Hilldale Court
Annandale, Virginia 20200
(703) 555-1492

November 15, 19XX

Memorandum to: All Staff

From: Jason Cartwright

Subject: Promotions

The company's reorganization has created three new positions, and I am pleased to announce the first two promotions into those positions.

First, Catherine Hair will become the new director of marketing. The design, sales, and marketing support activities will now report to her. As you know, Catherine has been with the company for a long time. Her background in both sales and advertising make her a natural choice for this position. Additionally, in the past she has taken the lead in the preliminary designs for our highly successful Tudor line of homes.

Next, Willis Sanders will assume the job as sales manager at Bay Hills and will also supervise the sales staff. Willis has been with us for only seven months and has already given Bay Hills a much needed jump start. That community was on dead zero when he took it over and last month it was our second hottest neighborhood. One of his first tasks will be to develop a sales training program that incorporates some of what he has done at Bay Hills.

Congratulations and good luck to Catherine and Willis in their new positions!

As you know, one position still remains open on the new organizational chart. I hope to make an announcement about this position very soon.

Finally, the newly promoted people described in Figure 4.6 will want to know exactly what is expected of them. So, you will need to carefully inform them of their new duties in writing as shown in Figure 4.7. However, if you describe their duties too thoroughly, you run the risk of stagnating creativity and inventiveness. It's a fine line, but you have to walk it.

Figure 4.7 Memo on New Job Duties

Jason Cartwright Builders, Inc.
101 Hilldale Court
Annandale, Virginia 20200
(703) 555-1492

November 15, 19XX

Memorandum to: Catherine Hair

From: Jason Cartwright

Subject: New Job Duties

As the recent reorganization continues to gel, we need to develop complete job descriptions and lists of job duties. In fact, one of your first tasks will be to develop these descriptions and duties for each position in your area, including your own. As we have discussed, I want everybody to know their job duties without limiting creativity and self-reliance that is so key to this business.

In the meantime, the following items will serve as a brief list of your job duties, subject to change as we continue to grow and expand:

1. The director of marketing will supervise the marketing support activities, sales managers, and designer.

2. The director of marketing will chair a committee formed by the sales managers and designer, which will develop and make recommendations for marketing strategies. These recommendations will include, but not be limited to, new home designs, new sales strategies, and advertising and promotional campaigns. The director of marketing will ensure input from appropriate people in the construction area for home design proposals.

3. The director of marketing will negotiate advertising and promotional contracts, subject to approval by the management committee.

Again, this list of job duties is just a start. We will need to describe them in greater detail in the near future. Hopefully, though, this list will help you to begin work in your new position.

In the end, the format of a job description and list of job duties will reflect the personality and style of the head of the company. Every employee, even the boss, wants to know where he or she stands. No one wants to make decisions not knowing whether he or she has the authority to make those decisions. The important thing to remember is that, as your staff grows and changes, you need to put something in writing about these changes, including reorganizations, promotions, and new job duties.

Expanding the Marketing Horizon into New Areas

Eventually, as your company grows, it will begin to change its marketing horizon. When you make those changes, even if they are gradual, you will have a lot of explaining to do—again, in writing. The memo and letters in Figures 4.8 through 4.12 are a start at handling those explanations.

Figure 4.8 Memo on a New Marketing Strategy

Jason Cartwright Builders, Inc.
101 Hilldale Court
Annandale, Virginia 20200
(703) 555-1492

March 15, 19XX

Memorandum to: All Employees

From: Jason Cartwright

Subject: New Marketing Strategy

The company has recently decided to begin a shift into southern Virginia and plans to open a branch office in Petersburg toward the end of the summer. As the company plans this move, we need the broadest possible input from all of the staff.

Initially we plan to increase staff only slightly in sales and construction supervision for southern Virginia. Any other additions to staff will come in a few years once we are on line in these two areas.

To start planning for this change, we need to take several steps. First, the personnel manager needs to assess the availability of high-quality employees and subcontractors in the affected areas.

At the same time, the accounting manager needs to examine the company's ability to service additional debt while we are bringing the proposed branch on line. The accounting manager also should determine the additional vehicle miles and possible cost of establishing a branch office.

Please bring your suggestions about this planned action to the next staff meeting or put them in a memo to me as soon as possible. As each of you know, we all share tangibly in the success of Jason Cartwright Builders. As with everything we do, this new direction in our marketing strategy will require teamwork from each of us to succeed.

If you plan to make a change in your marketing strategy like the one described in Figure 4.8, you will also need to notify brokers and clients of that change in writing. As shown in Figures 4.9 through 4.12, these letters should be personalized rather than form letters. At the same time, they can also reuse well-written text from earlier letters or memos. This situation is another example of the worth of a good word processor.

Figure 4.9 Letter to a Broker in an Existing Area

Jason Cartwright Builders, Inc.
101 Hilldale Court
Annandale, Virginia 20200
(703) 555-1492

August 17, 19XX

Ms. Carol Crawley
The Crawley Agency
Post Office Box 4
West Annandale, Virginia   20053

Dear Carol:

Through the years, The Crawley Agency, your agents, and clients
have maintained a strong, profitable relationship with Jason
Cartwright Builders. In the coming years, we want to see this
relationship strengthened as our company continues to grow.

Tomorrow Jason Cartwright Builders will announce the opening of
a new construction and marketing branch in Petersburg, Virginia,
to serve the southern Virginia market. This market presents some
excellent, new opportunities for the company. However, our
continued commitment to Annandale and the surrounding markets
also remains strong.

For years our company has built the finest custom homes in this
area, and we will continue to do so for years to come. We also
plan to continue our strong relationship with your agency. In
addition, I have a personal commitment to Annandale because it
is my home.

If you have any suggestions or questions about this planned
expansion, please give me a call. As always, I look forward to
hearing from you.

Sincerely,

Jason Cartwright
President

Figure 4.10 Letter to a Broker in a New Area

Jason Cartwright Builders, Inc.
101 Hilldale Court
Annandale, Virginia 20200
(703) 555-1492

August 17, 19XX

Mr. Lewis Mason
Mason Homes
4487 Richmond Pike
Petersburg, Virginia 23354

Dear Mr. Mason:

Tomorrow Jason Cartwright Builders will announce the opening of
a new construction and marketing branch in Petersburg to serve
the southern Virginia market. Your reputation as a leading
broker in the Petersburg area is well established, and we are
interested in developing a strong working partnership with a
broker who has a commitment to excellence that matches our own.

A Jason Cartwright home means quality and customer service
wherever it is built. Our commitment to our customers has led
the company to become one of the fastest growing builders on
both sides of the Potomac. We plan to make that same commitment
to your area.

Enclosed are copies of testimonial letters from some of our most
recent clients. Please feel free to distribute them among your
staff and customers. Also enclosed is a proposed marketing
brochure that we plan to use in the Petersburg area.

As a first step toward developing a strong relationship with
your company, we would like to get your input on our plans for
the Petersburg area. While we are bringing the new branch on
line, I will be in Petersburg at least two days a week. Next
week I will call you to schedule a meeting to discuss our plans
in detail. Until then I look forward to meeting with you.

Sincerely,

Jason Cartwright
President

Enclosures

Figure 4.11 Letter to Current Clients

Jason Cartwright Builders, Inc.
101 Hilldale Court
Annandale, Virginia 20200
(703) 555-1492

August 17, 19XX

Mr. and Mrs. Fritz von Trapp
449 Northspring Drive
Annandale, Virginia 20055

Dear Mr. and Mrs. von Trapp:

Tomorrow Jason Cartwright Builders will announce the opening of
a new construction and marketing branch in Petersburg to serve
the southern Virginia market.

Although the new branch offers us an exciting opportunity for
growth, it in no way diminishes the importance of our relationship with the von Trapp family. Our commitment to you and your
new home remains as strong as ever.

If you have any questions or concerns, please know that you can
always call me.

Very truly yours,

Jason Cartwright
President

Figure 4.12 Letter to Past Clients

Jason Cartwright Builders, Inc.
101 Hilldale Court
Annandale, Virginia 20200
(703) 555-1492

August 17, 19XX

Mr. and Mrs. Horace Jones
301 Bay Loop
Annandale, Virginia 20056

Dear Mr. and Mrs. Jones:

Tomorrow Jason Cartwright Builders will announce the opening of
a new construction and marketing branch in Petersburg to serve
the southern Virginia market. As the owner of a Jason Cartwright
home, you can make an important contribution to this planned
change.

Shortly after you moved into your new home, you completed a
brief survey form that described why you selected us as your
builders. Your input at the time was very helpful, and we
appreciated your advice and comments.

In light of our current changes, however, we would appreciate
your helping us with another brief survey. Using the enclosed
form, please look back on your relationship with us, particular-

```
ly after-the-sale service, and let us know whether there is
anything we need to improve.

Thank you for your help on this survey. We look forward to
hearing from you soon.

Sincerely,

Jason Cartwright
President

Enclosure
```

As the letters in Figures 4.9 through 4.12 demonstrate, you can reuse a core description of the marketing expansion in even the most customized letters to brokers and clients without too much difficulty.

Changing the Product Mix

As your business grows, you may change your product mix as well as expand your markets into new areas. Sometimes a shift in product mix can present greater challenges for your company than a shift in territory. However, with effective correspondence as shown in Figures 4.13 through 4.16, you can turn potential difficulties into advantages.

Figure 4.13 Invitation to a Party

Jason Cartwright Builders, Inc.
101 Hilldale Court
Annandale, Virginia 20200
(703) 555-1492

```
May 1, 19XX

Ms. Marianne Candy
Candy Realty
917 East Bayshore Boulevard
Annandale, Virginia  20057

Dear Marianne:

You are cordially invited to a reception on May 24 at 7:00 p.m.
at the Annandale Golf Club. That evening Jason Cartwright
Builders will unveil plans for a shift in product mix and
solicit input from you and other trusted business associates.

The company plans to continue as the area's leading builder of
upscale, custom homes. However, we also want to use our reputa-
tion for quality to take advantage of opportunities in the
move-up market.
```

Please give John Kelly in my office a call if you can join us on
May 24. I look forward to seeing you there.

Sincerely,

Jason Cartwright
President

Figure 4.14 Letter to a Supplier

Jason Cartwright Builders, Inc.
101 Hilldale Court
Annandale, Virginia 20200
(703) 555-1492

May 25, 19XX

Mr. Dominic DeAngelo
Courtesy Plumbing Company
Rural Route 4, Box 210
Annandale, Virginia 20059

Dear Dominic:

Thanks for joining us at the reception yesterday evening. We
greatly appreciated your comments on our plans to build move-up
homes and look forward to your ongoing support.

For years you and I have enjoyed a good working relationship and
personal friendship. As you know, our joint commitment to
quality has treated us both well through many up and down
markets.

However, move-up customers will be much more cost-sensitive
than the upscale customers we are used to. So, we need to start
thinking about affordable quality. In other words, Dominic,
let's sharpen our pencils.

In the next two weeks, we need some ballpark estimates from you
on those house plans we discussed last week. Also, can someone
from your company meet with our designer and work out the bugs
you noticed in the plans? Please have that person call Catherine
Hair, our director of marketing, who will schedule the meeting.

Thanks again for your comments last night, and I look forward to
hearing from you soon.

Sincerely,

Jason Cartwright
President

After the reception on May 24, Jason should send all the people who attended a
personal letter or note thanking them for coming and giving their suggestions. As
shown in Figure 4.14, you can include the thank you note in a normal business
letter.

Figure 4.15 Letter to a Broker

Jason Cartwright Builders, Inc.
101 Hilldale Court
Annandale, Virginia 20200
(703) 555-1492

May 25, 19XX

Ms. Angela Warren
Warren Realty
653 Bypass Road
Annandale, Virginia 20060

Dear Ms. Warren:

Your firm has a great reputation for quality and service in the move-up market, and our company wants to deal with firms that believe in those traits as strongly as we do.

As you know, for years Jason Cartwright Builders has built the finest upscale homes in Annandale. Today the company announced the creation of a new product line for the move-up market in Annandale called Coventry Homes by Jason Cartwright Builders. In this line, we plan to offer high-quality homes at an affordable price.

Although we have made a strong commitment to our existing brokers, most of them will admit that their strengths lie with a different market. That is where your firm comes in. With our mutual commitment to quality and service, our two companies can forge a profitable working relationship in the move-up market.

If you are interested, we would appreciate the opportunity to have you and some of your agents visit with us and discuss our plans for the move-up market in Annandale. Please call Catherine Hair, our director of marketing, to schedule a presentation for some of your key people. We look forward to hearing from you soon.

Sincerely,

Jason Cartwright
President

Figure 4.16 Letter to a Developer

Jason Cartwright Builders, Inc.
101 Hilldale Court
Annandale, Virginia 20200
(703) 555-1492

May 30, 19XX

Ms. Amelia Kent
Kent Properties
1100 Dixon Line Drive
Annandale, Virginia 20047

Dear Amelia:

Thank you for joining us at the reception last week. We greatly
appreciated your comments in person and by letter regarding our
plans to expand into the move-up market.

Please accept my assurances that Jason Cartwright Builders will
continue to build homes for our core market, the upscale customer. In fact, we believe that our image for quality in that
market will differentiate us in the move-up market.

As you know, the new product line for the move-up market will be
called Coventry Homes by Jason Cartwright Builders. This name
will allow us to capitalize on our reputation for quality in a
market that responds to such a reputation.

At the same time, to further cement our image in the upscale
market, we will strengthen the use of "Jason Cartwright Homes"
as a distinctive signature for top-of-the-line homes.

You and Kent Properties can rely on our company as a key builder
of high-quality homes in its communities for years to come. You
have my word on that. Thank you again for coming to our
reception and giving us your excellent comments last week.

Sincerely,

Jason Cartwright
President

Chapter 4 has provided some sample letters that a builder may need to use as his
or her company grows and expands. Certainly the experienced builder can cite other
situations when correspondence is required during business growth, but the examples in this chapter can guide the writing of those letters as well.

The next chapter looks at the expansion of a building company in a different
direction from expanding a staff or changing a marketing territory or product mix.
Chapter 5 focuses on expansion through diversification and shows the special
correspondence required in a few different ventures.

Letters for a Diversifying Business

Many home builders diversify at some point. Diversification can be *vertical*, such as a home builder opening a lumberyard. This kind of expansion occurs along a vertical line that leads from the raw material suppliers to the ultimate customers. A *horizontal* diversification is an expansion in the same business layer, such as a home builder opening a branch office, merging with another builder, or expanding into another segment of the building business.

This chapter does not try to present letters for every kind of diversification in building. Instead, it offers letters for three common areas into which a builder may diversify: light commercial building, remodeling, and property management. You can then adapt these sample letters to your own situation as it relates to diversification.

Expanding into Commercial Building

If you plan to expand your homebuilding business into commercial building, you will need to write letters to suppliers, lenders, and prospective clients as shown in Figures 5.1 through 5.3.

Figure 5.1 Letter to a Commercial Supplier

Jason Cartwright Builders, Inc.
101 Hilldale Court
Annandale, Virginia 20200
(703) 555-1492

January 5, 19X7

Mr. Melvin Lumas
Sales Manager
Southern Steel Supply Company
33 Marlboro Pike
Annandale, Virginia 20061

Dear Mr. Lumas:

Because your company has a reputation for quality and service, we would appreciate an opportunity to meet with you or a representative of your company to discuss opening an account.

As one of Annandale's leading builders of high-quality homes, Jason Cartwright Builders plans to expand into light commercial building this year. This expansion is part of a well-developed

growth plan established a few years ago with the help of our
lenders, suppliers, and brokers.

Growth in this direction will require dealing with several new
commercial suppliers. Because of your excellent reputation, we
are interested in working with your company as one of those
suppliers.

Enclosed is our most recent financial statement as well as a
list of our current suppliers. Please feel free to call any of
them for references. Also enclosed are letters of reference from
two of our lenders and two of our larger suppliers. Any one of
these people will gladly discuss our situation with you at
greater length.

If you require any other information from us, please give me a
call. I look forward to meeting with you or someone from your
company to discuss opening an account sometime soon.

Sincerely,

Jason Cartwright
President

Enclosures

Typically you will send letters such as the one in Figure 5.1 to several suppliers,
including many of Southern Steel's competitors. Again, this is an advantage of
owning a word processor. You can adapt the language in one letter to use in several
letters.

Your residential construction lender probably cannot make commercial construction loans. So, as a new commercial builder, you will probably want to establish a
banking relationship with the commercial lending department of a bank. You will
need the services of this department in a variety of situations, and a discussion with
them should probably precede any thought of expansion into commercial building.
However, first you will need to contact the department by letter to schedule such a
meeting, as shown in Figure 5.2.

Figure 5.2 Letter to a Commercial Building Lender

Jason Cartwright Builders, Inc.
101 Hilldale Court
Annandale, Virginia 20200
(703) 555-1492

October 22, 19X6

Ms. Julia Ruffin
Vice President and Manager
Commercial Lending Division
State Street Bank and Trust
3 Main Street Tower
Annandale, Virginia 20045

Dear Ms. Ruffin:

For several years now Jason Cartwright Builders has enjoyed a
strong and profitable relationship with your residential lending

division. We started with one construction loan in 19X2 and have
expanded to almost a third of our total construction financing
needs this year.

The company's growth plan, which is a carefully detailed agenda
of long-range needs, calls for an expansion into light commercial building. Our mutual friends in residential lending have
suggested that we meet to discuss the financing side of this
decision.

To prepare you for such a meeting, I have enclosed a five-year
financial statement as well as other documents that should give
you a clear picture of Jason Cartwright Builders. At your convenience, I would be happy to meet with you, either at your
office or mine, and look forward to hearing from you soon.

Sincerely,

Jason Cartwright
President

Enclosures

Finally, when diversifying into commercial building, you will have to deal with
that all-important first client. A diversifying business still has to throw off that
new-kid-on-the-block image, but that is easier to do with a positive reputation
behind you. That is also easier to do in a letter where you can polish up what you
say to the prospective client.

Figure 5.3 Letter to a Prospective Commercial Client

Jason Cartwright Builders, Inc.
101 Hilldale Court
Annandale, Virginia 20200
(703) 555-1492

March 12, 19X7

Dr. W. Johnson Smith
18 Wayfare Way
Annandale, Virginia 20063

Dear Dr. Smith:

In the past five years, Jason Cartwright Builders has construc-
ted some of the finest homes in Annandale. In your neighborhood
alone, Tarpon Run, we have built five homes in the past three
years. This year the company plans to bring that same commitment
to quality to commercial construction.

At Jason Cartwright Builders, we pride ourselves in maintaining
excellent relationships with our past clients. Enclosed is a
list of the five homeowners in Tarpon Run who have agreed to
provide you with references for our company. They have also
volunteered to show you their homes, if you so desire.

```
Please let us know if we can provide you with any additional
information about our company. Thank you for your interest in
Jason Cartwright Builders.

Sincerely,

Jason Cartwright
President
```

Figures 5.1 through 5.3 show just a few of the letters a home builder may need to write as he or she diversifies into light commercial building. Use them as models to craft other letters you may have to write as you develop your commercial building business.

Expanding into Remodeling

For many home builders, remodeling is a natural direction for diversification. In fact, many single-family builders got started as remodelers and never fully abandoned remodeling when they shifted into homebuilding.

Diversifying into remodeling has several distinct advantages for home builders. The most important one is that it is a countercyclical business. In other words, the remodeling business tends to be up when new home sales are down. Thus, remodeling is a good way for a homebuilding business to smooth out cash flow. However, managing a remodeling company requires some distinct talents and traits. The letters in Figures 5.4 through 5.10 give you a glimpse at some of the special challenges the remodeler faces.

Figure 5.4 Prospecting Letter for a New Remodeler

Jason Cartwright Builders, Inc.
101 Hilldale Court
Annandale, Virginia 20200
(703) 555-1492

```
January 5, 19X1

Mr. and Mrs. John de la Niece
88 Willow Lane
Annandale, Virginia  20064

Dear Mr. and Mrs. de la Niece:

Have you recently thought about replacing your old air condi-
tioning or heating system? Does the 50-year-old wiring or
plumbing in your home need a checkup? Perhaps you would like to
replace your drafty, single-pane windows with up-to-date
thermally efficient windows.

Jason Cartwright Builders is ready to help you with all of your
remodeling needs. Many of your neighbors have recently increased
the livability of their homes through prudent remodeling deci-
sions. Our company has the expertise to advise you on any
renovations, including for kitchens and baths. As you know, re-
modeling can add substantially to the resale value of your home
while improving your comfort today.
```

Our company provides warranties for our work and stands behind
everything we do. The company is licensed and bonded and is a
member of the local home builders association. In addition, we
provide references with every bid.

Please give me a call if we can assist you in your remodeling
decisions. I look forward to hearing from you.

Sincerely,

Jason Cartwright
President

As you can guess, the letter in Figure 5.4 would work for new as well as established remodelers. The letter in Figure 5.5 to the same couple capitalizes on the qualities that an established home builder possesses as he or she begins to diversify into remodeling.

Figure 5.5 Prospecting Letter for a Builder/Remodeler

Jason Cartwright Builders, Inc.
101 Hilldale Court
Annandale, Virginia 20200
(703) 555-1492

January 5, 19X1

Mr. and Mrs. John de la Niece
88 Willow Lane
Annandale, Virginia 20064

Dear Mr. and Mrs. de la Niece:

Have you recently thought about replacing your old air condi-
tioning or heating system? Does the 50-year-old wiring or
plumbing in your home need a checkup? Perhaps you would like to
replace your drafty, single-pane windows with up-to-date
thermally efficient windows.

Jason Cartwright Builders is one of Annandale's leading builders
of fine homes. Over the years we have built our reputation as
carefully as we have built our homes, and today our past clients
are our best references. This year we have expanded into remodeling and intend to deliver the same high quality and service to
satisfy your remodeling needs.

Many of your neighbors have recently improved the livability of
their homes through prudent remodeling decisions. Our company
has the expertise to advise you on any renovations, including
for kitchens and baths. As you know, remodeling can add substantially to the resale value of your home while improving your
comfort today.

Since we are a small, locally based business, I personally
supervise all of our jobs. We provide warranties for our work
and stand behind everything we do.

Please give me a call if we can assist you in your remodeling
decisions. You incur no cost or obligation for an initial
consultation. I look forward to hearing from you.

Sincerely,

Jason Cartwright
President

Old or new companies can use the *first* job in a particular neighborhood to generate more business. Every established remodeler can tell stories about starting a job, only to have neighbors from down the street drop by to solicit a bid on additional work. Putting a sign in the front yard of a remodeling job—if local zoning permits it—is a given. However, as the letter in Figure 5.6 demonstrates, the aggressive remodeler will go one step further.

Figure 5.6 Form Letter to Neighbors

Jason Cartwright Builders, Inc.
101 Hilldale Court
Annandale, Virginia 20200
(703) 555-1492

June 3, 19X1

Mr. and Mrs. Samuel Skinner
92 Willow Lane
Annandale, Virginia 20064

Dear Mr. and Mrs. Skinner:

Your neighbors, the de la Nieces, have chosen Jason Cartwright
Builders to renovate their home at 88 Willow Lane. Since we are
in your neighborhood, I would appreciate the chance to discuss
your own remodeling needs with you.

As you know, prudent remodeling can improve the livability of
your home. Our company has the expertise to advise you on any
renovations, including for kitchens and baths. Remodeling can
also add substantially to the resale value of your home while
improving your comfort today.

Since Jason Cartwright Builders is a small, locally based
business, I personally supervise all of our jobs. We provide
warranties for our work and stand behind everything we do.

Please give me a call if we can assist you in your remodeling
decisions. You incur no cost or obligation for an initial
consultation. I look forward to hearing from you.

Sincerely,

Jason Cartwright
President

Of course, if you had prospected on Willow Lane with a form letter at some earlier date, you would want to avoid the same wording in the letter shown in Figure 5.6.

Real estate agents can become excellent sources of remodeling leads. Many new homes require either pre- or postsale renovations, and agents will often need a ready remodeler who understands the need for moving quickly on these matters. As shown in Figure 5.7, you may want to use a letter to solicit remodeling leads from agents.

Figure 5.7 Letter to an Agent Regarding Pre- and Postclosing Repairs

Jason Cartwright Builders, Inc.
101 Hilldale Court
Annandale, Virginia 20200
(703) 555-1492

August 10, 19X1

Ms. Harriet Bowen
Bayside Realty
33 Market Square
Annandale, Virginia   20067

Dear Ms. Bowen:

Your company has established a reputation as one of Annandale's leading agents for older, upscale homes. Over the past several years, Jason Cartwright Builders has also established a reputation for building high-quality homes in the newer neighborhoods around the bay.

Now our company plans to bring that same commitment to quality into the remodeling arena. As a home builder who is used to meeting closing deadlines, we can offer you special insights on helping your clients make needed repairs immediately before or after closing.

If you agree, in the near future we would appreciate the opportunity to handle some of your must-do jobs. As an established builder with crews working every day, we can move people to the site almost immediately and bring the resources to finish the job on time.

Harriet, give us a chance to show you what we can do and you are sure to be pleased with the results.

Sincerely,

Jason Cartwright
President

With remodeling, you must *always* thank the customer after the job is finished as shown in Figure 5.8. And you must *always* ask the customer for a referral. That is the first law of selling remodeling. Future business will depend on getting those referrals, especially from past customers.

Figure 5.8 Letter Thanking Customers for a Job

Jason Cartwright Builders, Inc.
101 Hilldale Court
Annandale, Virginia 20200
(703) 555-1492

November 12, 19X1

Mr. and Mrs. John de la Niece
88 Willow Lane
Annandale, Virginia 20064

Dear Mr. and Mrs. de la Niece:

Thank you again for having Jason Cartwright Builders remodel
your kitchen and hall bath. Those new rooms are sure to bring
you years of enjoyment as well as added value to your home.

At Jason Cartwright Builders, we are always looking for
homeowners like you who need our remodeling services. Do you
know of anyone--perhaps some of your friends--who could use
those services? If you don't mind, I will call you in the next
few days to ask whether you have any referrals for us.

Again, thank you for the opportunity to remodel part of your
home. We look forward to working for you again if the need
arises.

Sincerely,

Jason Cartwright
President

Does calling the de la Nieces for a referral as suggested by the letter in Figure 5.8 seem too pushy to you? Prospecting for leads, if done correctly, does not have to be pushy. After all, your clients earn a living too. They understand that you have to sell your company's services. If you did a good job for your clients, they will appreciate the request and want to help you.

Publicity is another selling tool. Many homeowners do not like publicity, but many others relish it, particularly if it gives them a chance to brag about their newly remodeled home. Most large daily newspapers have someone in the position of real estate editor. This person will not become your personal public relations agent. However, when one of your jobs comes together in a special way, you will want to let the real estate editor know about it by letter, as shown in Figure 5.9.

Figure 5.9 Letter to a Real Estate Editor

Jason Cartwright Builders, Inc.
101 Hilldale Court
Annandale, Virginia 20200
(703) 555-1492

August 12, 19X1

Mr. Robert Lesly
Real Estate Editor
<u>The Annandale News</u>
Annandale, Virginia   20069

Dear Mr. Lesly:

Your piece on kitchen renovation last June was excellent. At
Jason Cartwright Builders, we give a copy of it to each of our
clients as an example of innovative thinking in kitchen reha-
bilitation.

Next month, we begin work on a similar renovation for a couple
in northern Annandale. In all my years of building and remodeling, this is by far the finest kitchen I have ever seen in-
stalled in Annandale.

If you are interested, the couple has agreed to meet with you
and show you what we have in mind. Enclosed are several sketches
that give you a rough idea of our plans. However, to get the
complete idea of what we intend to do, you need to visit the
site.

If you don't mind, I will call you in the next few days to
discuss this project. At that time, if you are interested, we
can schedule a site visit at your convenience.

Sincerely,

Jason Cartwright
President

Enclosures

The successful remodeler will live or die by referrals. You must *always* thank
people who give you referrals—preferably in writing as shown in Figure 5.10.

Figure 5.10 Letter Thanking a Customer for a Referral

Jason Cartwright Builders, Inc.
101 Hilldale Court
Annandale, Virginia 20200
(703) 555-1492

```
April 9, 19X1

Ms. Rhonda Simpson
101 Gibbs Street
Annandale, Virginia   20070

Dear Ms. Simpson:

Thank you for referring the Hustons to us. Referrals from valued
clients like you are especially appreciated. They are without
question the highest praise you can give us.

When we spoke last, you mentioned the possibility of our re-
modeling your upstairs bathroom. One of our plumbing suppliers
is having a close-out sale on bathroom fixtures. If you are
ready to make that move, now would be an excellent time to save
some money.

Again, thank you for the referral, and please feel free to call
on me whenever I can return the favor.

Sincerely,


Jason Cartwright
President
```

Of course, Ms. Simpson will probably call Jason in the near future. As Figure 5.10 demonstrates, never pass up an opportunity to sell yourself to past clients, even as you are thanking them for a referral.

Expanding into Property Management

For some builders another option for diversification is property management. This is particularly true for the builders who remodel and have some excess bookkeeping capacity. Naturally, diversifying into property management also requires some additional skills, training, and licensing. However, many builders and remodelers find it is a good way to smooth out revenue flows. The remodeling crew will have plenty of slack-time jobs rehabilitating old properties, maintaining current ones, and possibly even building new properties for existing investor-clients.

If you plan to enter property management, you will need to add a few more letters to your correspondence toolbox. Although the letters in Figures 5.11 through 5.14 do not cover every situation you will encounter in property management, you will find them helpful in a variety of situations.

Figure 5.11 Letter to a Prospective Tenant

Jason Cartwright Builders, Inc.
101 Hilldale Court
Annandale, Virginia 20200
(703) 555-1492

August 12, 19X1

Dr. Samuel Skinner
101 8th Street
Annandale, Virginia 20071

Dear Dr. Skinner:

Thank you for allowing me to show you the office space at 409 Cedar Terrace yesterday. This letter summarizes what we discussed at that time. Please review it carefully and call me if the summary is in error on any point.

1. The office is 2,400 gross square feet (gsf) to be rented at the rate of $10 per gsf the first year, $12 per gsf the second year, and $14 per gsf the third and final year of a three-year lease. A one-month security deposit will be required, and rent is due on the first of each month.

2. The tenant will be responsible for all utility payments. The owner will be responsible for all taxes and assessments, except garbage service.

3. The office will be remodeled at the owner's expense according to the enclosed specifications, at a total cost not to exceed $3,000.

Also enclosed is a sample of our rental contract for you to review. If you are in agreement with the terms of the contract or have any other points you wish to discuss, please call me.

Thank you again, and I look forward to hearing from you soon.

Sincerely,

Jason Cartwright
President

Enclosures

Often rental agencies will refer tenants, particularly commercial tenants, to you. Local real estate associations usually have guidelines for commission splits or referral fees in these situations. If not, as shown in Figure 5.12, you should make your company policy clear on this issue from the start.

Figure 5.12 Letter to a Rental Agency

Jason Cartwright Builders, Inc.
101 Hilldale Court
Annandale, Virginia 20200
(703) 555-1492

August 12, 19X1

Ms. Jennifer Corley
Corley Properties
The Annandale Building, Suite 100
Annandale, Virginia 20072

Dear Ms. Corley:

Thank you for referring Dr. Samuel Skinner to us. I showed him
the property at 409 Cedar Terrace yesterday and he seems interested in it. The owners, who live in Jamaica, have given us some
latitude in negotiating with tenants, so I think we can meet his
requests.

In addition, this letter summarizes our discussion of the
referral fee on the telephone last week. For referrals on
one-year leases, we pay one-half of the first month's rent,
which is somewhat higher than the local average.

For referrals on multiple-year leases, we pay one-half of the
first month's rent times the number of years of the lease. So,
for example, if Dr. Skinner signs a three-year lease, he will
pay $2,000 per month the first year (2,400 gross square feet
times $10 divided by 12 months). You then receive one-half of
that amount or $1,000 times three years or $3,000 as your
commission.

Furthermore, we pay the commission over a number of months equal
to the number of years of the lease. Therefore, we will send you
a fee of $3,000, divided into three equal monthly payments, upon
receipt of each of Dr. Skinner's first three monthly rental
payments.

If you have any further questions about the referral fee, please
give us a call. Thank you again for the referral, and we look
forward to doing business with you again in the future.

Sincerely,

Jason Cartwright
President

Believe it or not, some tenants are occasionally late with their rent. You need to
be very tactful about this. Clearly, you are always a salesperson. If tenants have
legitimate cash-flow problems, then you can become their friend by working with
them. As shown in Figure 5.13, a letter to a tenant about late rent must achieve a
balance between tact and firmness.

On the other hand, if you have a problem tenant on your hands, local or state
landlord-tenant laws will severely limit your ability to bring—or even threaten—
legal action. A whole body of local, state, and federal court rulings and the senti-

ments of your local courts will also limit your ability to bring or threaten legal action. When in doubt, and there is plenty of room for doubt on this issue, always contact qualified local advisers for guidance.

Figure 5.13 Letter to a Tenant Regarding Late Rent

Jason Cartwright Builders, Inc.
101 Hilldale Court
Annandale, Virginia 20200
(703) 555-1492

```
May 17, 19X2

Dr. Samuel Skinner
409 Cedar Terrace
Annandale, Virginia  20073

Dear Dr. Skinner:

We regret to see a good tenant like you fall behind on your rent
payments. However, your account is now three months past due.
Furthermore, several telephone calls from our office to yours
have produced no response.

If there is anything we can do to assist you in this situation,
please let us know and we will make every effort to do so. For
example, the property owners have given us the authority to
restructure your past-due amount and spread it over several
months if that will help. They have also authorized me to
renegotiate the rental agreement to give you some more time to
catch up in your payments if necessary.

In addition, please let us know if you have a concern about the
service or condition of the building. To date, you have not
indicated to us that you have such concerns and our most recent
inspection has revealed no problems. Furthermore, local land-
lord-tenant laws require that rent be paid during any disagree-
ment over the service or condition of the building.

Dr. Skinner, you are just the kind of tenant we want for that
property, so please let us know if we can help you in any way to
resolve this matter.

Sincerely,

Jason Cartwright
President
```

Finally, just as with remodelers and commercial builders, property managers need to solicit business. As you diversify into other building-related areas like remodeling, commercial building, and property management, you should never miss a chance to prospect for business! The letter in Figure 5.14 shows you one way to do that in property management.

Figure 5.14 Letter Seeking Referrals from Neighbors

Jason Cartwright Builders, Inc.
101 Hilldale Court
Annandale, Virginia 20200
(703) 555-1492

October 10, 19X1

Ms. Edith Bromley
Bromley Investment Securities, Inc.
Middletown Office Mall, Suite 5
Annandale, Virginia 20074

Dear Ms. Bromley:

As you know, Jason Cartwright Builders has become the new property manager for Middletown Office Mall. As a full-service
building and remodeling company, we plan to begin improvements
at Middletown as soon as possible. In the next few weeks, I will
visit each tenant to solicit your input and advice on the
planned changes.

As with any business, Middletown Office Mall can only succeed
with a steady supply of clients. Although the mall has several
vacancies right now, they will not impede the planned improvements in any way.

However, additional tenants of the same quality as Bromley
Investment Securities would enhance the vitality of Middletown
as a place to conduct business. We would therefore appreciate
any referrals of potential tenants that you might give us.

Thank you for your contributions to the planned renovations at
Middletown, and I look forward to meeting with you soon.

Sincerely,

Jason Cartwright
President

Chapter 5 has provided sample letters that a builder might use when diversifying into light commercial construction, remodeling, or property management. The chapter does not cover every situation you will face as you begin to diversify your business. However, you can tailor the sample letters in this chapter to the particular situations you encounter. You can also apply the principle "When in doubt, put it in writing!" to many of the situations involved in diversification.

Having covered the correspondence required for diversifying your business in Chapter 5, the book will now present in Chapter 6 the correspondence required for one of the most important areas of your business: customer service.

Customer Service Letters

Every letter you write in the building business is in some way a customer service letter. This book has often cited examples of general customer service letters you may write in the normal course of building, marketing, and closing homes. Chapter 6 offers some more specific customer service correspondence. It also shows ways to use this correspondence to address customer service issues you may face each day in your business.

General Customer Service Letters

While building a home, you may need to review several items with or obtain certain decisions from customers. You will find putting such information or requests in writing beneficial. As shown in Figure 6.1, you may need to be helpful in your letter but also firmly encourage the customer to take some action.

Figure 6.1 "It's Time to Pick out Your Carpet!" Letter

Jason Cartwright Builders, Inc.
101 Hilldale Court
Annandale, Virginia 20200
(703) 555-1492

February 14, 19X1

Mr. and Mrs. Gus Carpenter
St. Julian Place, #12
Annandale, Virginia 20075

Dear Mr. and Mrs. Carpenter:

In the next two weeks, we will enter the finish stage of your new home. My assistant, Susan Kirkland, tells me that you have yet to make your finish selections. To complete your home on schedule, we need your final selections as soon as possible.

At your earliest convenience, please provide Susan with your final finish selections for the following six items:

1. ceramic tile from the Tile Center in Fenway Plaza

2. interior paint and stain for the office (We have the samples here.)

3. vinyl flooring for the kitchen and utility room from Custom Carpets on South Otis Street

4.   carpet also from Custom Carpets

5.   bath and kitchen hardware from Sonny's Hardware, also in
Fenway Plaza

6.   appliance colors also from Sonny's Hardware

Susan is available to help you with your selections if you need
her. She will check with you in a few days to see whether she
can assist you in making these selections.

Thank you, and I look forward to seeing you again soon.

Sincerely,

Jason Cartwright
President

If a customer requests a change order while you are building a home, you may
want to send a cover letter with the change order form to the customer, as shown in
Figure 6.2. As part of that letter, you may want to explain the change order process
to the customer.

Figure 6.2 Cover Letter for a Change Order

Jason Cartwright Builders, Inc.
101 Hilldale Court
Annandale, Virginia 20200
(703) 555-1492

August 12, 19X1

Ms. Sally Jones
855 Dewey Avenue
Annandale, Virginia 20076

Dear Ms. Jones:

The remodeling of your kitchen is moving along, and we are
pleased that you are satisfied with the progress on it so far.
Your new kitchen is sure to bring you many years of pleasure as
well as increase the value of your home.

As we discussed yesterday, enclosed are preliminary sketches for
remodeling the downstairs half-bath. Also enclosed is the change
order that authorizes this work.

As the change order indicates, the price for remodeling the
half-bath is quite reasonable. If we start immediately, we can
probably finish the bath along with the kitchen.

Please study the change order carefully. If you and Mr. Jones
agree, sign and return one copy to us so we may begin work on
remodeling the half-bath.

Thank you again for this opportunity to be of service to you.

Sincerely,

Jason Cartwright
President

Enclosures

Sometimes construction jobs run over schedule. While such delays are often unavoidable, they can quickly frustrate your customers. As soon as you see a possible delay coming, you need to notify them in writing as shown in Figure 6.3. Keep your apologies to the customers short and simple.

Figure 6.3 Letter on Scheduling Delays

Jason Cartwright Builders, Inc.
101 Hilldale Court
Annandale, Virginia 20200
(703) 555-1492

March 1, 19XX

Ms. Sally Jones
855 Dewey Avenue
Annandale, Virginia 20076

Dear Ms. Jones:

As you know, building your new custom-designed kitchen cabinets has required considerable millwork in the shop. This in turn has caused a delay in finishing your new kitchen.

As a standard business practice, we build delay time into each job schedule, but our estimate in this case was too conservative. As a result, rather than an April 30 finish on your home, we are now looking at a May 31 completion.

Jason Cartwright Builders deeply regrets this delay. We pride ourselves on keeping clients satisfied. Our reputation in this respect is most important to us, and we work hard to maintain it. If you need our assistance in any way, please let us know.

Thank you for your patience. We will finish your home as soon as possible.

Very truly yours,

Jason Cartwright
President

The letter in Figure 6.1 dealt with a situation where the builder needed to coax a customer to make final finish selections. The letter in Figure 6.4 addresses another type of customer delay that you may face. This letter also takes a polite but firm approach similar to the letter in Figure 6.1. In writing letters to customers, remember to carefully select a tone for your letter that is appropriate for the given situation.

Figure 6.4 Letter to Encourage a Customer's Performance

Jason Cartwright Builders, Inc.
101 Hilldale Court
Annandale, Virginia 20200
(703) 555-1492

May 3, 19X3

Mr. and Mrs. Howard Holmes
The Shire, Apartment B
Annandale, Virginia 20078

Dear Mr. and Mrs. Holmes:

As you recall, your contract requires that you apply for a
mortgage loan within 10 days after signing the contract and
inform us of your selection of a lender at that time.

Since we signed the contract a few weeks ago and have not yet
heard from you on this matter, we wondered whether you had
forgotten to notify us of your selection of a lender. If you
need another copy of your contract for your lender or any other
information that we can provide, please give us a call.

If we have not heard from you in a few days, we will call to see
whether you need any additional information on your contract for
your lender. Thank you.

Sincerely,

Jason Cartwright
President

As shown in Figure 6.4, you should try to take a pleasant, helpful approach when
encouraging customers to perform. In these situations, try to avoid developing an
adversarial tone. If you use such a tone, you run the risk of upsetting the customer
and preventing the very performance you are trying to encourage with your letter.
Besides encouraging customers to perform, you often need to coax performance
from brokers and agents as well. Figure 6.5 gives an example of an occasion when
you might need to encourage a broker to perform.

Figure 6.5 Letter to Encourage a Broker's Performance

Jason Cartwright Builders, Inc.
101 Hilldale Court
Annandale, Virginia 20200
(703) 555-1492

```
May 3, 19X3

Glynnis Franks
Franks Homes
The Bayshore Building
Annandale, Virginia   20079

Dear Glynnis:

You and I have worked closely together for many years. During
that time your company has demonstrated professionalism, exper-
tise, and determination in handling the sales of our homes.
Together we now need to turn our attention to the Holmes' con-
tract.

As you know, the Holmes are buying 108 Berkeley. Your agent,
Jack Barlow, is handling their contract, which we signed several
weeks ago. Since we have not yet heard from them regarding the
selection of their mortgage lender, we were wondering whether
Jack has heard any further information on this matter.

If you wouldn't mind, I would like to call you in a few days to
find out whether the Holmes are having second thoughts on their
contract and we need to stop construction of their home. Togeth-
er, you and I should be able to resolve this issue and hopefully
keep the job on track. Thanks for your help on this.

Sincerely,

Jason Cartwright
President
```

Writing a letter to a broker about a buyer who has delayed in applying for a mortgage loan is quite appropriate. However, notice how Jason opens his letter in Figure 6.5 by making a few positive statements to Glynnis about their ongoing working relationship. He then takes a helpful and inquiring rather than adversarial approach with her in trying to find out more information about the delayed mortgage application.

Jason carefully avoids directly criticizing the agent who is handling the contract. In addition, throughout the letter and especially at the end, he uses words like "together" to emphasize a team approach to resolving the situation. As a result, Glynnis will probably respond to this letter in a positive manner and take some appropriate action.

Mortgage lenders traditionally deal directly with the customer before closing, and many builders are happy letting a real estate agent handle all of those details. Sometimes, though, you need to step in and assist in the process. In fact, you may need to communicate with the lender in writing as shown in Figure 6.6. Although you have a reasonable right to some information, you have to watch your step here because confidentiality rules limit what the lender can tell you. So, make sure the tone of your preclosing letter to a mortgage lender is tactful and appropriate.

Figure 6.6 Preclosing Letter to a Mortgage Lender

Jason Cartwright Builders, Inc.
101 Hilldale Court
Annandale, Virginia 20200
(703) 555-1492

```
August 12, 19X3

Ms. Janice Clifton
The Mortgage House
Post Office Box 8818
Annandale, Virginia  20080

Dear Ms. Clifton:

Thank you for your note regarding the mortgage application for
the Holmes, who are purchasing the property at 108 Berkeley. We
appreciate your expert handling of this mortgage.

Let me mention just a few points that may assist you in prepar-
ing the closing packet:

*  Our attorney, Marvin Gilstrap, can prepare the deed less
expensively than your closing attorney, so you may want to use
him.

*  Burton and Debreau, PE, performed the preliminary survey, so
you should contact them for the final survey.

*  We have all the property tax receipts in this office, as well
as the municipal garbage and water contracts and some other
pertinent documents. Please have the closing agent contact us
for them at the appropriate time.

*  We will prepare the home warranty on the day before closing
and bring this document to closing.

Next Tuesday, Wednesday, and Thursday are the days that I am
available for closing. Please contact my assistant, Susan
Kirkland, once you have picked the time and date for closing.
Thank you, and we look forward to doing business with you again
in the near future.

Sincerely,

Jason Cartwright
President
```

Occasionally you may need to write a preclosing letter to a building inspector about final inspection. For example, a customer's last-minute change order request may or may not affect the final building inspection. As you know, building inspection varies widely from locale to locale. Some municipalities have exacting building inspection departments with a myriad of rules and requirements specific to that city.

At the same time, a rural area just a few miles away may seem to have by comparison hardly any inspection system at all.

Many builders deal with inspectors totally by telephone and in person. In other situations or places, you may find that written communications with the inspector about final inspection are in order. The letter in Figure 6.7 is an example of this situation and may or may not apply to your locale. If you are a new builder or new to an area, you may want to ask the advice of a more experienced builder in your area on this issue. As an alternative, you can also discuss this issue with the head of your local home builders association.

In any event, you certainly do not want to face delays in closing and thus risk upsetting your customer because of a glitch in the final building inspection. So, you should carefully craft any letter about final inspection to a building inspector. Also, keep in mind that you establish a reputation for professionalism through all of your business letters, including those sent to building inspectors.

Figure 6.7 Preclosing Letter to a Building Inspector

Jason Cartwright Builders, Inc.
101 Hilldale Court
Annandale, Virginia 20200
(703) 555-1492

August 12, 19X3

Mr. Bobby Hazard
Chief Building Inspector
The City of Annandale
Annandale, Virginia 20081

RE: Building Permit B-12289

Dear Mr. Hazard:

Within the next two weeks, we will complete the residence at 108 Berkeley Way and will call you for a final inspection at that time.

In reviewing our file on this project, we note that we made a minor modification to the air-handling system to accommodate a change in the heating system at the customers' request. (See the enclosed amended plans.) This change also produced some minor changes to the wiring plan. Although these changes did not require an amendment to the building permit, we wanted to let you know about them before the final inspection.

Your staff has been extremely helpful on this project. As you know, most of our work is done in the suburbs, so we rarely get a chance to deal with your office. However, all of us at Jason Cartwright Builders look forward to having an opportunity to work with your office again in the future.

Sincerely,

Jason Cartwright
President

Enclosures

Every builder, even the finest, has occasional complaints from customers. As shown in Figure 6.8, the more professional builders respond to these complaints immediately and in writing. A letter responding to a customer's complaint impresses even the hardest person that you are truly interested in his or her concerns. This kind of letter also leaves a paper trail of the steps you took to resolve the issue in case the situation ever goes really sour.

Figure 6.8 Letter Responding to a Customer's Complaint

Jason Cartwright Builders, Inc.
101 Hilldale Court
Annandale, Virginia 20200
(703) 555-1492

August 15, 19X3

Mr. and Mrs. Howard Holmes
The Shire, Apartment B
Annandale, Virginia 20078

Dear Mr. and Mrs. Holmes:

Thank you for sending me a list of your concerns about your new home. Whenever clients have concerns, particularly this close to the move-in date, we want to deal with them immediately.

The following list shows each of your concerns and indicates the steps we have taken or will take to remedy them before closing:

1. Squeaky kitchen floor--The subflooring was improperly nailed. We have corrected this.

2. Ceramic tile in the master bath--We agree that the grouting job fell beneath our standards and will regrout this floor tile.

3. Front-yard drainage--We still have some final landscaping to do and intend to finish it on the day before closing.

4. Wrong light fixture in the dining room--Our electrician has replaced this light fixture.

In addition, thank you for bringing to our attention your concern over the color of the paint in the hall bathroom. After checking our records, we discovered that we have painted the hall bathroom the color that you selected--robins egg blue. Enclosed is a copy of the signed work order and the color swatch that you used to make your selection.

As we have mentioned to you, a color will often appear different on the wall from the way it appears on the small swatch. If you would like the hall bathroom repainted, we will gladly accommodate you. Please sign and return the enclosed change order, which indicates the charge for this service, and we will begin work immediately.

On the day before closing, we want to meet with you at the house and walk through it carefully. During the walk-through, we will use a checklist that has proven valuable in spotting warranty items. The walk-through should take about two hours. My assis-

```
tant, Susan Kirkland, will call you soon to schedule a time for
the walk-through.

Thank you again for sharing your concerns with us. We will make
every effort to address any remaining concerns before closing.

Sincerely,

Jason Cartwright
President

Enclosures
```

Remember that some of the most important letters you will ever write in your business are customer service letters, especially letters responding to a customer's complaint. Craft the wording of these letters carefully. Many customers' concerns are legitimate, and you will need to address them if you want any referrals from these customers.

On the other hand, some customers' concerns may not be legitimate. In those cases, as shown in paragraphs three and four of the letter in Figure 6.8, you will need to clearly state in your letter to the customer why those concerns are not legitimate. Therefore, in most customer service letters, a pleasant but firm tone is usually appropriate. For a summary of tips on writing effective customer service letters, see Figure 6.9. In fact, you can apply the tips in Figure 6.9 to all of your business correspondence.

Figure 6.9 Tips on Writing Effective Customer Service Letters

1. Open and close letters with a compliment or positive statement.
2. Be helpful but firm.
3. Make factual statements and back them up.
4. Avoid overusing "I" and "we."
5. Stay neutral and businesslike. Avoid emotions.
6. Focus on the issues.
7. Avoid an adversarial tone.
8. Emphasize teamwork to address the situation.

Warranty Letters

Good builders have their own detailed warranty and customer service programs and budget into each job a warranty cost based on past experience. Customers appreciate this kind of attention. Also, anticipating these problems and developing a system for handling them will save you substantially even in the short run. The warranty letters in Figures 6.10 through 6.14 are examples only. You will need to adapt them to your own warranty, customer service, or insurance programs if you have them.

If you do not have a warranty or customer service program in place, you may find these letters helpful in developing such a program. In addition, two books published by Home Builder Press of the National Association of Home Builders provide detailed information on how to develop warranty and customer service programs: *Customer Service for Home Builders* by Carol Smith and William Young and *Warranty Service for Builders and Remodelers* also by Carol Smith.

The first warranty letter you may need to write is a preclosing letter as shown in Figure 6.10. Among other things, this letter needs to review procedures for the customer-and-builder walk-through just before closing and provide a brief description of your formal warranty program.

Figure 6.10 Preclosing Warranty Letter

Jason Cartwright Builders, Inc.
101 Hilldale Court
Annandale, Virginia 20200
(703) 555-1492

```
August 24, 19X3

Mr. and Mrs. Howard Holmes
The Shire, Apartment B
Annandale, Virginia  20078

Dear Mr. and Mrs. Holmes:

Thank you again for buying a Jason Cartwright home. As you know,
your new home is now finished. As closing approaches, we wanted
to explain how Jason Cartwright Builders will assist you during
the closing process.

Our company is a long-standing member of the Southeastern War-
ranty Program (SWP). This broad-based program provides 10 years
of structural limited warranty on your home. At closing we will
provide you with the necessary insurance papers and other docu-
ments on this program.

In addition, we have an in-house customer service program that
goes beyond SWP. We are proud of our commitment to service, and
over the years our customers have been uniformly pleased with
this program. As a first step in the program, we will ask you
to meet with us next Monday, the day before closing, to walk
through your new home.

During the walk-through, we will discuss every component of your
new home in detail and will go point by point over our 10-page
checklist--often called a punchlist--to ensure that every item
in your home meets or exceeds your expectations. Then we will
both sign the checklist, and you will keep one copy of the
signed checklist for your records.

If anything is noted on the checklist to be fixed, our construc-
tion supervisor will try to follow up on each agreed-upon item
within 30 days. Within a week after the walk-through, the con-
struction supervisor will make an appointment for our workers to
visit your home and correct any items on the list that day. How-
ever, if any items on the list require hiring subcontractors or
ordering parts and material, the work may take a bit longer. At
that time, we will give you a completion schedule for any final
work.
```

Thank you again for buying a home from Jason Cartwright
Builders. We look forward to seeing you next Monday for the
walk-through of your new home.

Sincerely,

Jason Cartwright
President

Although you sent the letter in Figure 6.10 to the customers *before* closing, you
will probably reexplain your warranty and customer service programs to the
customers *at* closing. In addition, you will need to send them another letter *after*
closing to explain your customer service procedures again as shown in Figure 6.11.

Figure 6.11 Postclosing Warranty Letter

Jason Cartwright Builders, Inc.
101 Hilldale Court
Annandale, Virginia 20200
(703) 555-1492

August 30, 19X3

Mr. and Mrs. Howard Holmes
108 Berkeley Way
Annandale, Virginia 20084

Dear Mr. and Mrs. Holmes:

Congratulations on moving into your new home! All of us at Jason
Cartwright Builders wish you many happy years there.

As part of our customer service program, I have reviewed the
checklist from the walk-through with our construction supervisor, Billy Joe Spears. He will call you in the next few days
to make an appointment for our workers to visit your home.

At that time, we intend to complete every item on the list.
However, a few items may take longer if they require hiring
subcontractors or ordering parts and material. Mr. Spears will
discuss this with you when he visits your home.

Thank you again, and I look forward to seeing you soon.

Sincerely,

Jason Cartwright
President

cc: Billy Joe Spears

Truly excellent companies in every industry believe that the customer always
comes first. Many studies and books have been written on the subject, and customer-oriented companies consistently come out ahead in their fields. This does not
mean, however, that the customer will always get everything he or she wants. It
does mean, though, that you will be clearly aware of and sensitive to customers'
feelings every time you have to disappoint them. The letters in Figures 6.12 through
6.14 deal with the same warranty situation and show three different outcomes. As

you read these three letters, notice how their tone and wording change depending on the circumstances.

Figure 6.12 Warranty-Issue Letter #1

Jason Cartwright Builders, Inc.
101 Hilldale Court
Annandale, Virginia 20200
(703) 555-1492

October 10, 19X3

Mr. and Mrs. Howard Holmes
108 Berkeley Way
Annandale, Virginia 20084

Dear Mr. and Mrs. Holmes:

Thank you again for bringing your concerns over the drainage of your home to our attention. As you recall, you first raised this issue right before closing, and we responded to it in a letter dated August 15, 19X3. At the time, I reminded you that we would perform some additional landscaping before closing and this should satisfy your concerns over the drainage.

On the day before yesterday I visited your home, but unfortunately you were not there. That day I took some photographs of the site and have enclosed copies of them. The spots where you indicated some concerns about the drainage are circled with grease pencil on the photos.

Bobby Hazard, the chief building inspector for the City of Annandale, was with me when I visited your home and gave me his informal opinion on the matter. He agrees that the landscaping as it now exists is more than adequate to handle the natural drainage needs of your property. He also put that opinion in a letter to me. Enclosed is a copy of his letter for your review.

In addition, he and I both noticed that you had placed your lawn sprinkler very close to the edge of your house (see photo #2). If you are using this sprinkler to water your shrubs, this could cause some dampness under your house in that area. Unfortunately, we can do nothing for you in that situation.

If you have any other questions or we can help you further in this matter, please let us know.

Sincerely,

Jason Cartwright
President

Enclosures

Figure 6.13 Warranty-Issue Letter #2

Jason Cartwright Builders, Inc.
101 Hilldale Court
Annandale, Virginia 20200
(703) 555-1492

October 10, 19X3

Mr. and Mrs. Howard Holmes
108 Berkeley Way
Annandale, Virginia 20084

Dear Mr. and Mrs. Holmes:

Thank you again for bringing your concerns over the drainage of
your home to our attention. As you recall, you first raised this
issue right before closing, and we responded to it in a letter
dated August 15, 19X3. At the time, I suggested that we would
perform some additional landscaping before closing and this
should remedy any potential drainage problems.

On the day before yesterday I visited your home, but unfortunately you were not there. That day I took some photographs of
the site and have enclosed copies of them. The spots where you
indicated some concerns about the drainage are circled with
grease pencil on the photos.

As you can see from the photos, the final landscaping was
performed improperly and this is probably causing the drainage
problem. To correct the problem, we will need to remove the
shrubbery and relandscape the areas indicated in the photos.

We sincerely regret that this problem was not caught earlier.
Please accept my personal apologies for this oversight. Naturally, our company will bear all expenses associated with this
work. In addition, we will try to finish the relandscaping
before the end of the month.

I will stay in touch with you until we have finished the relandscaping. If you have any questions or we can help you in any
way, please let me know.

Sincerely,

Jason Cartwright
President

Enclosures

Figure 6.14 Warranty-Issue Letter #3

Jason Cartwright Builders, Inc.
101 Hilldale Court
Annandale, Virginia 20200
(703) 555-1492

October 10, 19X3

Mr. and Mrs. Howard Holmes
108 Berkeley Way
Annandale, Virginia 20084

Dear Mr. and Mrs. Holmes:

Thank you again for bringing your concerns over the drainage of
your home to our attention as quickly as you did. This early
notice will allow us to resolve the problem before winter comes.

On the day before yesterday I visited your home, but unfortunately you were not there. That day I took some photographs of
the site and have enclosed copies of them. The spots where you
indicated some concerns about the drainage are circled with
grease pencil on the photos.

As you recall, you first raised this issue right before closing,
and we responded to it in a letter dated August 15, 19X3. At the
time, I suggested that we would perform some additional landscaping before closing and this should remedy any potential
drainage problems.

After that preclosing work was completed, we followed up by
asking two drainage consultants to review the matter and also
solicited the opinion of Mr. Bobby Hazard, the chief building
inspector for the City of Annandale. We all agree that an
underground drainage system is needed immediately in the areas
indicated on the enclosed photos.

Although part of the front yard needs to be removed to create
this system, we can save much of the fresh sod and new shrubbery
that is now in place. Furthermore, our company will replace any
sod and shrubbery we cannot save. With your written permission,
this work can begin at once and hopefully be completed before
the holidays.

I will stay in touch with you until we have corrected this
matter. If you have any other questions or we can help you in
any way, please let me know.

Sincerely,

Jason Cartwright
President

Enclosures

Chapter 6 has presented examples of letters you can use to address a few common customer service issues you may face as a builder. Again, these letters do not cover every customer service situation, so you will need to adapt them to your own circumstances. However, remember this important point: If you have a customer service issue to resolve, you will benefit greatly from putting your response to the customer in writing. By doing this you create an invaluable record of what goes on between you and the customer. More importantly, when customers receive such a letter, they will appreciate the written proof of your concern for their concerns.

Chapter 7

Letters for Challenging and Special Situations

After reading the title of this chapter, many builders may say, "I don't have anything *but* challenging situations!" Actually, that is probably true. After all, if the building business were easy, everyone would want to enter it. However, some situations in building are more challenging than others, and so this last chapter leaves you with a few ideas for dealing with those special situations that you may face from time to time.

Letters for Business Reversals

Every business person sometimes faces a business reversal. If you take the initiative to deal with the reversal, you increase your chances of surviving it unscathed. As shown in Figures 7.1 through 7.3, a few well-worded letters to key business associates, such as lenders, suppliers, and brokers, will go a long way toward helping you weather a tough period.

Figure 7.1 Letter to a Construction Lender

Jason Cartwright Builders, Inc.
101 Hilldale Court
Annandale, Virginia 20200
(703) 555-1492

March 12, 19X9

Mr. Milburn Douglas
President
Commerce Bank of Annandale
Annandale, Virginia 20009

Dear Mr. Douglas:

Enclosed is a copy of our most recent audited financial statement for your files. As you can see, right now Jason Cartwright Builders faces a financially challenging period.

Over the years our company has built an excellent reputation in Annandale. Our record for customer service, integrity in dealing with suppliers, lenders, and employees, and careful, conservative business planning is well known and admired.

Two extraordinary situations have coincided to put us in our current situation. First, as you know, the recession has hit the local homebuilding industry hard. In response to this, we have expanded into remodeling to smooth out our revenue sources.

However, some unanticipated start-up costs in this area have
further reduced our cash flow. Nonetheless, remodeling is
expected to become a strong source of cash for the company by
the year's end.

Also, as a custom home builder, we are in a sensitive position.
Most of our jobs are presold. However, if a job falls through,
we are left holding a custom home that typically does not sell
as quickly as a production home. In 19X8 three large jobs fell
through, and by year's end, we had only sold one of them. These
jobs carry a large interest expense, as indicated on the financial statement. However, we are now taking additional steps to
sell the other two homes and reduce that interest expense.

Jason Cartwright Builders remains a strong, healthy company at
the core in spite of these financial challenges and will continue to be one of your best customers in the future. Thank you
for your patience and willingness to see us through these challenging times. Together hopefully we can look forward to a
return to prosperity sometime soon.

Sincerely,

Jason Cartwright
President

Enclosure

Suppliers will often be more aware of your circumstances on a daily basis than
lenders. After all, they are in the same specific industry, see their own sales receipts
fall off during the same business cycles, and tend to communicate with you more
often. As a result, rather than transmitting news, a business reversal letter to a
supplier will discuss how the two of you can work out of the problem together, as
shown in Figure 7.2.

Figure 7.2 Letter to a Supplier

Jason Cartwright Builders, Inc.
101 Hilldale Court
Annandale, Virginia 20200
(703) 555-1492

March 12, 19X9

Mr. John Arnold
President
Arnold Lumber Company
8895 River Rapids Road
Annandale, Virginia 20031

Dear John:

You and I have done business with each other for many years, and
this is the first time Jason Cartwright Builders has had a
past-due situation with your company. As we recently discussed,
our company currently faces a financially challenging period,
but we are making every effort to address our problems head on.

As you know, to smooth out revenue sources during the current
recession, the company has expanded into remodeling. However,
some unanticipated start-up costs have further reduced our cash
flow. Nonetheless, remodeling is expected to become a strong
source of cash for us by the year's end.

Also, as you are well aware, three large jobs fell through in
19X8, and by the year's end, we had only sold one of them. These
jobs carry a large interest expense for the company. However, we
are now taking additional steps to sell the other two homes and
reduce that interest expense.

Jason Cartwright Builders remains a strong, healthy company in
spite of these financial challenges and will continue to be one
of your best customers in the future. Right now, though, we need
your patience and willingness to see us through these challenging times.

Enclosed is a copy of our most recent audited financial statement. This gives you a clear picture of where we are and where
we need to go. After you have had time to review the statement,
I will call to make an appointment to see you. Perhaps together
we can find an effective way to handle our accounts with your
company under the circumstances.

Thank you again for your help and understanding during this
challenging period.

Sincerely,

Jason Cartwright
President

Enclosures

Finally, you are going to need help working your way out of any business
reversal. Old business associates will have suggestions, experiences, and insights
they can share with you. Communicating your needs to them in writing is a first step
in that direction, as shown in the letter to a broker in Figure 7.3.

Figure 7.3 Letter to a Broker

Jason Cartwright Builders, Inc.
101 Hilldale Court
Annandale, Virginia 20200
(703) 555-1492

March 12, 19X9

Ms. Anne Baxter
President
Baxter Realty
101 Bayside Boulevard
Annandale, Virginia 20035

Dear Anne:

For years your firm has been a leading seller of Jason Cartwright homes, and we have always valued your insight and

experience in this challenging business. In addition, your company's hard work and perseverance in selling the home on Cedar Creek Drive last December was very much appreciated.

As you know, right now our company has two other homes that remain unsold. When you have a moment, we would really appreciate your suggestions on how to move these homes.

If you don't mind, I will call you next week to schedule a meeting. Perhaps by putting our heads together, we can come up with some ideas that will sell these homes soon. Thank you for your ongoing support.

Sincerely,

Jason Cartwright
President

Again, notice how Jason describes the business reversal in the letters to a lender, supplier, and broker in Figures 7.1 through 7.3. He tries to emphasize the excellent reputation his company has established over the years and the positive steps he has taken to stop the reversal. This upbeat approach will make an important impression on the people receiving these letters.

Letters Dealing with Subcontractors and Suppliers

Chapters 2 through 5 presented letters that dealt with run-of-the-mill issues a builder might face in relationships with suppliers and subcontractors. Figures 7.4 and 7.5 present letters that deal with more serious matters that may arise in those relationships. As always, when faced with these kind of situations, respond to them in writing in order to create a record. Also clearly indicate the nature of your concerns and the steps you have taken or plan to take to deal with them.

Figure 7.4 A Mild Complaint Letter

Jason Cartwright Builders, Inc.
101 Hilldale Court
Annandale, Virginia 20200
(703) 555-1492

June 5, 19X3

Mr. Joshua Porter
Porter's Framing and Drywall Company
Rural Route 2, Box 818
Annandale, Virginia 20091

Dear Joshua:

You and I have worked together on many jobs in recent months, and your company has done good work for us on those jobs. However, a situation came up yesterday that we need to discuss.

During lunch our construction superintendent, Billy Joe Spears, found several of your workers visibly drunk on our job site at 101 Highgate Road. Several empty beer cans were scattered around

the site and the supervisor on the job, although apparently
sober himself, was fully tolerant of the situation.

Mr. Spears rightfully asked the crew to leave the job site and
not come back. He then took several photos of the beer cans
before cleaning up the site. Copies of these pictures are
enclosed.

As past letters to you have stated, our company has a firm
policy about not drinking on the job. If any employee or
subcontracted worker breaks this rule, that is grounds for
dismissal. There are no exceptions.

Your company has done good work for us in the past, and we
think you can continue to do good work for us in the future.
Please give me a call to schedule a meeting, so together we can
find a way to prevent this situation from happening again. Thank
you for your help on this.

Sincerely,

Jason Cartwright
President

Enclosures

Once again notice how Jason begins the complaint letter to the subcontractor
in Figure 7.4 with a positive statement and ends on an upbeat note. At the same
time, Jason clearly states in factual terms what has happened, what was done about
it, and where he and Joshua need to go from here to resolve the situation. At the end,
he also suggests that, if he and Joshua work together on the situation as members of
the same team, together they may be able to address it more successfully. In letters
like the ones shown in Figures 7.4 and 7.5, you may find that using a matter-of-fact
rather than adversarial tone will help you more quickly to address the situation and
move on.

Figure 7.5 A More Serious Complaint Letter

Jason Cartwright Builders, Inc.
101 Hilldale Court
Annandale, Virginia 20200
(703) 555-1492

January 10, 19X7

Mr. Ted Smithson
President
Smithson Lumber Supply Company
Post Office Box 408
Annandale, Virginia 20092

Dear Mr. Smithson:

On November 15, 19X6, we began ordering all of our framing
material from your company. However, since December 15, 19X6,
our framing crew on the job site at 814 Beaver Creek Road has
consistently reported shortages in material.

```
This morning, the construction superintendent and I checked your
delivery ticket for framing lumber for 814 Beaver Creek Road
against the material actually on the truck. We both counted the
load and came up with the same total, which is $623.18 (before
tax) less than the amount on your delivery ticket. Enclosed is
an itemized list of material on the truck and a copy of the
delivery ticket.

Tomorrow I am meeting with our framing crew and bookkeeper to
estimate the accumulated shortages in your company's deliveries
since December 15. This should be relatively simple to do, and
we will give you a copy of the necessary supporting documents
for our estimate. Once you receive these documents and estimate,
please deduct the estimated amount from our account with your
company.

Thank you, and we look forward to hearing from you sometime soon.

Sincerely,

Jason Cartwright
President

Enclosures
```

As Figure 7.5 demonstrates, sometimes a direct approach is best in a letter that describes a more serious complaint you might have with a supplier or subcontractor. Even in this kind of letter, though, you still need to use a businesslike tone throughout and explain the situation as clearly as possible.

Letters to Building Permit Officials

Across the country the process for obtaining building permits varies from "pay your money and pick up your permit" to the more stringent and lengthy approaches often used in larger metropolitan areas. In many areas, you will regularly have to appeal the decision of a building permit, inspection, zoning, or business license official. The letter in Figure 7.6 offers an example of such an appeal.

Figure 7.6 Letter Appealing a Permit Decision

Jason Cartwright Builders, Inc.
101 Hilldale Court
Annandale, Virginia 20200
(703) 555-1492

October 1, 19X7

Dr. Julia Winstead
Planning Director
City of Annandale
Post Office Box 3
Annandale, Virginia 20093

Dear Dr. Winstead:

With this letter, we are appealing your ruling of September 25,
19X7, on the issuance of a building permit for a single-family
residence proposed for 108 Southwood Drive.

Please add this permit issue to the agenda for the next planning
and zoning committee meeting on October 10, 19X7. If possible,
please also have copies of this letter distributed to members of
the committee before that meeting.

According to your letter of September 25, this permit is denied
based on City Ordinance 2518.9, paragraph 7. That paragraph
requires that all single-family detached homes in zoning area 3R
be built on lots with a minimum street frontage of 70 feet. The
lot at 108 Southwood Drive has 68 feet of frontage.

As you know, the neighborhood on Southwood Drive is very old.
Given the available city records, we estimate that the lots were
originally laid out in the 1890s. As such, our lot clearly falls
within the spirit of the grandfather clause outlined in paragraph 16 of Ordinance 2518.9.

An additional two feet of frontage cannot be purchased from
either adjacent lots without disturbing the structures located
on them. However, the property owners of those lots have signed
letters of support for the construction of this new residence.
Those letters are enclosed.

Also enclosed are letters from two appraisers attesting that the
highest and best use for this parcel, located in an established
residential neighborhood, is as a residential building lot.

In addition, we have enclosed an artist's rendering and floor
plan of the proposed structure and a surveyor's drawing of the
proposed building site with the footprint of the structure
traced in. These drawings show that a creative architectural
design has removed any aesthetic issues associated with a
68-foot frontage.

We also have buyers lined up for this residence. These people
want nothing more than to move their growing family into this
well-established neighborhood, become good citizens of
Annandale, and add to the tax base.

```
For these reasons, we are asking for a waiver, as permitted
under paragraph 12 of Ordinance 2518.9, so that a building
permit may be issued for the residence proposed for 108
Southwood Drive. Thank you.

Sincerely,

Jason Cartwright
President

Enclosures (7)
```

In writing letters to building officials like the one shown in Figure 7.6, again you need to state your request in a clear and businesslike fashion and provide sufficient documentation to support your request. If you think your letter carries legal implications for your company, always have it reviewed by a qualified attorney.

Guidelines for Letters to the Editor

Writing a letter to the editor is a good, old American tradition. It is also an effective communications and public relations tool. Business leaders and associates, customers, and the general public all read letters to the editor on a regular basis. If you become known as an effective communicator, your local home builders association may even ask you to write an occasional letter on behalf of your fellow builders and associates.

Newspaper editors receive many letters every day. They reserve the right to edit these letters, to publish only segments of them, or to toss them in the trash. If you want to get your letter to the editor published, make sure you get your message across with these few, simple guidelines:

1. Keep your letter short. Most editors specify a maximum length—200 words is not uncommon. Much shorter is better.
2. Stay on the subject. Pick one topic and just talk about it. Rambling letters get tossed in the can.
3. Address ideas, not people. Even if a public official does something you dislike, address the *actions* of the official, not the official.
4. Avoid overused expressions, often called clichés.
5. Avoid racial, sexual, or ethnic slurs.
6. Remember that you are a leader in this industry. When you sign that letter to the editor, you will be identified with all of the home builders in your area and they will be identified with you. Take that responsibility seriously.

Conclusion

Good communicators are not born. They educate themselves through careful and frequent practice. In other words, good letter writers write lots of letters. Only after sustained practice can you turn your letters into useful tools for your building business.

And what useful tools they are! Effective letters are powerful selling and public relations tools. Use them to impress your clients, agents, and brokers with your ability to communicate your ideas to them. Letters are also powerful negotiating tools. They allow you to set the tone and agenda for negotiations even before they start. They can save you time by communicating your needs and wants to others in clear, concise terms.

Letters are also powerful financing tools. As the financing side of homebuilding becomes more complicated, well-written letters can improve your ability to present yourself and your company in the most professional way to lenders. Letters and memos are also powerful management tools. They allow you to put your company policies down on paper for all to see, so no misunderstanding can arise among your employees, subcontractors, and suppliers over exactly what you mean.

However, in the end, letters and memos are just like any other tools. If you leave them in your toolbox rather than put them to work building your business, they are useless. Business leaders are occasionally polled on the traits most needed for success in their industry. Consistently across all industries *communications skills* rank number one. So, to succeed financially in your building business, make letters and memos your number one business tools today.